PSYCHOLOGY OF SPORTS

Dr. Sinku Kumar Singh
M.P.E., M.Phil, NET, Ph.D. (FELLOW)

KHEL SAHITYA KENDRA
7/26, Ansari Road, Darya Ganj,
New Delhi-110002

Published by:

KHEL SAHITYA KENDRA

7/26, Ansari Road, Darya Ganj, New Delhi-110002.

Ph.: (O) 011-65190343, 43551324, 42564726
(M) 9811088729, (Fax) 011-42564726
(Res.) 47091605

E-mail: *khelsahitya1@rediffmail.com*

Web: *www.khelsahitya.com*

Revised & Updated Edition

ISBN - 978-81-7524-543-3

PRINTED IN INDIA 2016

Laser Typeset by:
JAIN MEDIA GRAPHICS
Delhi

Printed by:
ASIAN OFFSET PRINTERS
Delhi

Price: Rs. 800/-

Contents

1

INTRODUCTION TO SPORT PSYCHOLOGY

Within the past few decades, a new field of sport science has been recognized. It is called sport psychology, and it is concerned with both the psychological factors that influence participation and performance in sport and exercise and the psychological study motivation, personality, aggression and violence, leadership, group dynamics, exercise and psychological well-being, thoughts and feelings of athletes, and many other dimensions of participation in sport psychology classes, conduct research, and work with athletes, coaches, and exercise participants to help improve performance and enhance the quality of the sport and exercise experience.

Coaches were interested in the psychological aspects of athletic competition even before there was a science called sport psychology. For example, in the 1920s Knute Rockne, the football coach of the fighting Irish of Notre Dame, popularized the pep talk by making it an important part of his coaching.

We should note, however, that Rockne did not attempt to psych up his team for every contest. He only used the pep talk for special occasions. Coaching interest in contemporary sport psychology also involves more than a mere concern for psyching up athletes for competition.

Applied sport psychology focuses on only one facet of sport psychology, that of identifying and understanding psychological theories and techniques that can be applied to sport and exercise to enhance the performance and personal growth of athletes and physical activity participants. Applied sport psychologists are part of a complex interrelationship between psychological theory, application, historical context, social influence, and the individual behaviour of the applier and the recipient.

The area of applied sport psychology has gone tremendously in recent years, as evidenced by the numbers of coaches and athletes now looking to sport psychology for a competitive edge.

These individuals are turning to various psychological training programmes to learn, among other things, ways to manage competitive stress, control concentration, improve confidence, and increase communication skills and team harmony. The goal of psychological training is to learn to consistently create the ideal mental climate that unleashes those physical skills that allow athletes to perform at their best.

Most of the early sport psychology research focused on attempts to link personality or character with participation in athletics. As such, the majority of those early investigations were designed to compare the personalities of athletes with those of non-athletes. When few definitive trends emerged from personality-based research, sport psychologists became disheartened with that approach and began to argue that a better avenue for understanding psychological aspects of individual sport behaviour might be theory testing. Throughout most of the 1970s sport psychologists searched the literature for theories in the parent discipline of psychology that might also be applicable to motor skills.

A good example of such theory borrowing is the use of social facilitation theory, which provided testable predictions of how the presence of other people should affect both initial skill acquisition and later skill performance. The most fundamental prediction was that the presence of other persons should be disruptive to early skill acquisition but facilitative to later performance. Much of this theory testing took place in laboratory settings, and the early laboratory-based results were promising in terms of developing sport psychology as a scientific field. Unfortunately, when those same predictions were taken to field settings for reality testing, little was uncovered in terms of practical applications to sport behaviour. As a result of the difficulties in applying theories such as social facilitation to sport, theory testing of this nature was almost totally abandoned by sport psychologists in the late 1970s and early 1980s.

Researchers began to ask more clinical questions such as "How does mental practice or imagery affect sport performance?" "How can relaxation or biofeedback assist in the management of sport-related stress?" Many researchers also studied coach-athlete compatibility and the relationship between team cohesion and team outcome. As a consequence of this more clinical approach to sport psychology, the field has moved closer to answering the questions posed most frequently by coaches.

This is certainly desirable; however, a distinct vacuum in sport psychology theory still exists. When sport psychologist rejected theory testing in favour of more humanistic concerns they also almost eliminated any systematic attempts at building theoretical understanding of sport-related behaviour. Such theory building is critical to providing a sound basis on which to draw the methods and strategies used to intervene or change behaviour within sport settings. On the other hand, many of today's leading researchers in the field are again arguing for and actively engaging in research that can lead to development of new theories that are unique to individual sport behaviour.

Research is taking place on the playing field, in the gymnasium, and in the laboratory . Questions are being answered for both practical and theoretical reasons. It is hoped that the next decade will continue to see much more sport-specific theory building as the discipline of sport psychology matures as a science in its own right, ultimately complete with its own theories and its own sound clinical principles.

PRINCIPLES OF SCIENTIFIC INQUIRY

To more clearly understand how a science develops and why theory development is so important, it is necessary to understand certain aspects of science and scientific inquiry. If sport psychology is to be considered a legitimate science, then it must use the methods and vocabulary common to all sciences. Throughout the text you will encounter common scientific vocabulary. We hope that by explaining the vocabulary, methods, and strategies of scientific inquiry, you will be free to concentrate on the pertinent topics and issues presented in the rest of this text. The unit of analysis in sport psychology is individual behaviour as it occurs in sport situations. To achieve an understanding of that behaviour, four steps must be followed.

Description and Explanation

First, you must attempt to describe exactly how a person behaves in an indefectible situation. For instance, you might observe that when other people are watching, the strokes of a particular tennis player become stiff and erratic. Once you see that this behaviour is rather consistent, the second step is to try to explain why such effects occur.

Therefore, once you are able to describe behaviour, you then try to explain the causes underlying that behaviour. Although the latter may seem relatively easy, in reality it is often most difficult. In our tennis example it is not enough to say that presence of spectators causes the negative effects on the tennis player's strokes. That merely describes what take place.

To explain more fully the relationship between the presence of spectators and negative effects on tennis performance, you can then test other hypotheses. For instance, instead of measuring the patterns of muscle activity, you can assess what is happening to the person's processing of environmental cues. Perhaps the tennis player is distracted by the presence of other persons and therefore splits his or her attention between watching the tennis ball and assessing what is happening among the spectators. It seems reasonable that such attentional scanning or switching can cause performance problems. Perhaps the player cannot see the oncoming ball soon enough and therefore hurries the stroke. It is through this process of testing alternative hypotheses to determine their "fit" that you can ultimately construct a sound explanation of the relationship between the presence of other persons and its effects on one person's tennis strokes.

Prediction and Intervention

The third step is a prediction. A primary goal of the science of sport psychology is the ability to predict individual sport behaviour under identifiable conditions. Returning to our tennis example, if nothing changes, you can reasonably predict that any time this tennis player attempts to play in the presence of spectators, his or her performance will be hindered. Should you simply stop at that point? Certainly not! Once you are able to make fairly definitive predictions, you can then apply principles of psychology so that behaviour can be changed or controlled. This is the process of intervention, the use of psychological principles to enhance individual sport performance. In the case of the tennis player, performance can be enhanced by applying any one of several intervention strategies. For example, the athlete can be instructed in the use of relaxation procedures so that control over muscular tension is increased. Or attentional control can be gained by teaching the performer to use verbal cues such as "bounce" and "hit" at the appropriate time. Sport psychology has the same goals as does any other behavioural science—the description, explanation, prediction, and change of individual behaviour.

Its uniqueness lies in its attempt to achieve those goals within the framework of sport and motor performance situations. The attainment of those goals allows you to apply strategies to assist the performer in enhancing sport performance.

IMPLICATIONS FOR SPORT

An effective coach or teacher must take a philosophical position on how people acquire, maintain, and improve their motor skills. This presents three models of somewhat differing views of that process. By adopting one of them, you can then gear your instructional strategies accordingly. For instance, if you adopt the external causation model, then you should concentrate on varying external factors such as stimulus conditions present before movement and the nature of the reinforcements provided after a response. If you favour the internal mediational model, then you should concentrate on the amount and content of the feedback you provide performers.

The steps of the scientific method provide a logical basis on which any teacher or coach can obtain the information needed to solve a problem. Use those steps to gather the information you may also find that you have become more effective as a problem solver. You should now have a fairly good grasp of many of the concepts and variables associated with undertaking scientific investigations in sport psychology.

FUNDAMENTAL TYPES OF RESEARCH

The research undertaken in the field of sport psychology involves (1) the relative applicability of the findings and (2) the origin of the hypothesis.

Quasi-Experimental Research versus True Experimental Research

To conclude our discussion of the scientific dimensions of the psychology of sport behaviour, one final notion needs to be addressed: the distinction between quasi-experimental research and true experimental research. Earlier we said that the independent variable in an investigations was considered to be the cause of some effect on a dependent variable. One example was the possible effect of failure on subsequent persistence behaviour at the same task. In that example, failure would be categorized as the independent variable and persistence would be classified as the dependent variable.

The distinction between quasi-and true experimental research rests on two criteria. The first is whether an independent variable is actually varied or manipulated in that particular study. The second criterion is whether there is a control group. A control group is a group of subjects who are not exposed to the independent variable. If the independent variable is not manipulated and there is no control group, the study is called a quasi-experiment.

If an independent variable is manipulated and there is a control group, then the study is considered a true-experiment. It is important to note that in either case some measurement is made on the dependent variable. Consider a situation in which an experiment notices that highly skilled gymnasts appear to exhibit lower degree of anxiety as compared with less-skilled participants. An anxiety test is then administered to both groups of gymnasts, and the data analysis reveals that the experimenter's hypothesis was correct.

On the average the more highly skilled athletes had lower levels of anxiety as compared with the less-skilled gymnasts. In this example a dependent variable was assessed. However, the independent variable was not actually varied by the investigator; the athletes possessed that skill before the study. In addition, there was no control group, such as a comparison group of non-athletes.

Therefore even though the study proposed a hypothesis, systematically collected data, and used appropriate statistical analysis, it would not be considered a true experimental design but rather a quasi-experiment. Suppose that another person observed that beginning tennis player seemed to perform better or worse as a consequence of the information that was provided to them after a response.

In terms of the content of the feedback, it appeared that beginners performed better after positive feedback and worse after negative feedback. An experiment was then designed in which half of the players performed and always received negative feedback, whereas the other half performed an identical task but always received positive feedback. After the collection and analysis of the data, no differences between the groups were shown. Was this a true or a quasi-experiment? Could it be considered to be a true experiment in that the independent variable was manipulated? Half the players were provided with negative feedback, whereas the other half received positive feedback.

However, something is missing from the experiment the design did not include a control group. In this case, to be a true experiment, a third group should have been included, one to which no feedback was provided. There are some very real advantages to designing true experiments.

The inclusion of a manipulated independent variable and a control group allows researchers to show cause -and-effect relationships. Those cause-and-effect relationships allows a science to move toward adequate explanations and eventually to the point at which genuine theory building can begin.

However, good quasi-experiments or descriptions of variables must usually precede the design and implementation of true experiments. The science of sport psychology is young and exciting. It is not limited to one experimental site or to a single research approach. It is equally appropriate to conduct studies in locations such as a swimming pool, a baseball diamond, or an experimental laboratory. Scientific inquiries are limited only by the creativity and imagination of the researcher, who can be a full-time researcher, a teacher, a coach, or an athlete.

Empirical Research versus Theoretical Research

The second dimension along which research in sport psychology can be conceptualized is the origin of the hypothesis to be tested. We used the term empiricism earlier to refer to the method of observation. Many research hypotheses are generated on the basis of the observations or experiences of the scientist. For example, a scientist might observe that there appears to be something different about the personalities of athletes who prefer team sports to individual sports.

This observation could lead to the development of a study designed to test the validity of that empirically based hypothesis. At the other end of the empirical dimension lies the element of the theoretically based hypothesis. In that instance, the hypothesis is derived directly from an existing theory depicts the two dimensions along which research can be classified. To better understand the interrelationships among the elements of applied, basic, empirical, and theoretical research, suppose that a particular scientist was interested in the question of how images are stored in memory.

Suppose further this interest was purely to gain new knowledge and that there was no available theory to guide the investigator. This investigation would be classified as being from the basic-empirical perspective. If, on the other hand, there was a theory available that predicted that images are stored in analog form the same study should be categorized as being within the domain of basic-theoretical research. From a different point of view, suppose that a teacher observed that some students seem to prefer and to excel in competitive situation and others seem to prefer and to perform better in cooperative situations.

A study could then be designed to ascertain exactly how the two groups of students might differ in terms of some psychological characteristic. The purpose of such an investigation might be to apply the results in a manner capable of changing the behaviour of one or the other of the groups.

In this case the researcher would be operating within the framework of applied-empirical research. It is interesting to note, however, that if that same teacher had been aware of achievement motivation theory, he or she would have known that theory would predict that persons characterized as being high in achievement motivation should prefer and perform well in competitive settings. On the other hand, persons who are assessed as being low in achievement motivation should prefer and perform best in cooperative situation. Therefore if a hypothesis were derived from the perspective of achievement motivation, and if the results were to be applied in a real-life situation, such as a class in physical education, then that investigation would be considered to fall within the category of applied-theoretical research.

Applied Research versus Basic Research

Research that is completed with an eye toward directly applying the results to a real-life situation is called applied research. Evidence obtained from such investigations is applicable to athletes in actual sport settings. Obviously, this is the type of research that most coaches and teachers want sport psychologists to do. They want answers to questions about psychological aspects of athlete behaviour. Thus applied research hold an extremely important position within the research efforts of many sport psychologists. Other researchers are more interested in basic research, in which they gather information concerning individual psychology in movement settings simply for the sake of increasing knowledge; they are not interested in the immediate application of their results.

Often students, teachers, and coaches do not understand why basic research is important because it has not clear-cut relationship to real-life behaviours or situations. Perhaps a contemporary example will help you understand why basic research is necessary. Few of us could deny the impact that contemporary high-technology information and applications have on our daily life.

What any persons do not realize, however, is that the fundamental mathematical principles and theoretical notions associated with those technological advances were generated from basic research; that is, the information was generated simply for the sake of the information itself. It was only later that other researchers took the basic information and applied it for various technological purposes. For instance, in the field of biomechanics, much of the research uses computer-assisted techniques. However, when those techniques were first developed, no thought was given to their application to movement analysis.

BASIC SCIENTIFIC CONCEPTS AND VOCABULARY

Ways of Knowing

To more fully understand sport psychology as a science, it is necessary to understand the basic concepts and vocabulary of scientific inquiry. In this first section we will discuss various ways in which scientist answer questions such as "How do we know that is truth?" "What is reality or fact?"

Phenomenology

How does a scientific know that something is true, that an observed event is real and factual? One way is to call on personal experience, or phenomenology. Many of the beliefs we all hold are based on our own experiences. In the field of sport psychology, much of what takes place when coaches attempt to apply psychological principles to improve the performance of athletes is based on their own past experiences. They reason that "if it worked for me, then it should work for the other person." However, this is not necessarily true;

people differ too much from one another. Therefore the phenomenological method is not considered adequate in a scientific sense. For a method to be adequate scientifically, it must allow for the generation of cause-and-effect relationships or laws that can be applied to large numbers of persons or events.

Authority

Another method of "knowing" that is used with great frequency in sport psychology is that of authority. From this perspective, one look to a major authority in the field to provide the answers or facts. For instance, many basketball coaches have looked to John Wooden, former UCLA basketball coach, for advice on how to best coach basketball. What's wrong with this approach? On the surface it seems logical to look to outstanding coaches for ideas on how to coach. But think for a moment: would Wooden's approach work equally well with high school students in a small low town? Would Pat Head Summit, coach of the U.S. Women's Olympic basketball team, be as successful as she was with the 1984 Olympic team if she used the same methods with a junior varsity basketball team? Is there a single, best approach to the coaching of all basketball teams, regardless of factors such as skill level, personality of the head coach, or gender of the players? Probably not. Thus the method of authority lacks sufficient scientific credibility because it fails to provide principles that can be generally applied.

Theory Building

The fourth method of knowing is theory building. A theory is developed by pulling together the known facts relative to a certain phenomenon and establishing predictions concerning how each may be related to the others. Those predictions are then subjected to testing through studies that follow the steps of the scientific method. Therefore, although the basic scientific procedures are identical to those followed by the empirical method, the purpose differs. In theory testing the hypothesis is derived directly from the theory, and the conclusion derived from the data analysis feeds back to the theory. In other words the conclusion indicates whether the prediction made by the theory was accurate. Therefore both the development and the eventual testing of theories is an efficient, ongoing process.

Developing a field of scientific inquiry requires both the method of empiricism and the method of theory building. Obviously, empirically based investigations must precede theoretically founded studies. A researcher needs a collection of facts before he or she can speculate on the interrelationships among those facts. Without empiricism there would be less possibility of generating fresh and innovative ideas. As you saw earlier in our example of what a coach would need to know about anxiety and athletic performance, without attempts at theory building a researcher would have only a cumbersome group of facts.

Theories, Models, and Paradigms

Frequently the three terms, model and paradigm are used synonymously. Actually they are quite different. A theory attempts to synthesize what is known about a collection of facts and to make predictions concerning the relationships among those facts. A theory can be conceived of as a collection of hypotheses that attempts to predict or explain the relationships between multiple independent and dependent variables.

Therefore a theoretical framework provides testable predictions concerning the interrelationships among the variables associated with a particular phenomenon. Although we realize that such a description of theory may leave you puzzled at this point, the concept will become clearer as we discuss several theories later in the text, such as theories concerning how anxiety should affect motor performance in predictable ways and conflicting theories concerning phenomena such as observational learning, imagery effects, and personality.

As you will see, all these theories have certain commonalities. They all identify independent and dependent variables of interest and they all propose specific hypotheses concerning the interrelationships among those variables. A model is an analogy that is devised to help you conceptualize something about the interrelationships between a set of variables. It is usually visual or graphic. When coaches use Xs and Os, that is a model of the play to be executed. The Xs and Os represent, or are analogous to, the actual players. When architects develop blueprints to represent the structure of a building or when pharmaceutical companies design plastic human figures to show how certain drugs such as aspiring move through the body, they have constructed models.

Scientists interested in the relationship between psychology and sport and motor behaviour have developed several models of their own. One of the earliest of those models was developed by Lewin. As shown below

$$B = f(P \times S)$$

this model simply indicates that behaviour is a function of the interaction between some characteristic of the person and some event in the situation. From this perspective, behaviour is seen as being determined by the interrelationships among elements contained within a certain person and the circumstances that are observable in the social situation. An alternative explanation for human behaviour is that proposed by the external causation model:

Stimulus—Response—Positive Reinforcement

It states that a human response is caused by some observable, antencedent environmental condition called a stimulus. Is such a response is followed by some pleasurable consequence labelled positive reinforcement, then that same or a highly similar response will tend to recur under those same stimulus conditions. The underlying notions of this model are that responses are generated on the basis of some external factor in the first place and are maintained by some other external. To better understand the external causation model, suppose that a Little League player is at bat with a single runner on third base. With no previous instructions or experience, the player bunts the ball. The coach is ecstatic and praises the athlete profusely. The external causation model would predict that the next time this player is at bat under those same stimulus conditions, the same response will most likely recur.

In this model the term stimulus is used to describe the sensory input into the human system, whereas cognition is used to describe how information is processed before eliciting a response. Therefore this model portrays humans as thinking beings who process environmental information before deciding on an appropriate behavioural response. For instance, this model would predict that an athlete would mentally process feedback or information from a coach before attempting to changes in those skills. Once a model is developed, it influences how a research experiment is designed.

The manner in which such an investigation is carried out is called a paradigm. A paradigm describes the experimental protocol or methodological procedures used in a particular investigation. These procedures are drawn logically from the theoretical framework, or model, on which the study is based.

We will use the external causation model and the internal mediational model to help clarify this definition. Suppose that two scholars were both interested in understanding how a person learns motor skills. If you were to walk unannounced into each of their laboratories, you would probably not find the same experiment underway because they probably would not have selected the same motor task or the same subjects.

However, for the sake of our argument, assume that they were both assessing motor skill acquisition on the same task by highly similar subjects. Would they be following identical procedures? Probably not. However, it is certain that each would be setting up an experimental arrangement that would be consistent with the theoretical model that each experimenter believed best explains the acquisition of motor skills.

If the first experimenter believed that such behaviours are acquired as a function of some external factor he or she would then logically design an experiment in which some factor in the environment was varied, such as whether positive reinforcement was provided after each response.

On the other hand, if the second scientific took the perspective that motor skills are learned primarily through cognitive processing then it would follow that he or she would construct an experimental protocol in which some aspect of cognition was varied.

Thus it is clear that the paradigm employed by each of the two scientific is drawn logically from the different theoretical perspectives they hold. Teachers and coaches follow the same pattern as scientific. If coach A believes that players can improve athletic performance primarily through means such as awarding letters or posting track times, then that coach is following the principles of the external causation model. If Coach B attempts to improve athletes' sport performance by pointing out errors and video-taping plays, then Coach B is trying to enhance performance by providing information to the players and is therefore applying the principles of the internal mediational model. Although the examples of experiments cited earlier both took place in a laboratory setting, there is no reason why similar studies cannot be conducted in field situations. The type and location of research studies in sport psychology are quite flexible.

Empiricism

The third approach to knowing is empiricism, which simple means " based on observation." Thus it is similar to the phenomenological method, which is based personal experience. However, it differs phenomenology in that empiricism refers to observations that are made under strictly controlled experimental conditions. From this perspective, a "fact" is generated by following the steps of the scientific method. The scientific method is common to all scientific inquiries, regardless of whether the question concerns physics, biology, sociology, or sport psychology. These are four basic steps in the scientific method.

First, a statement of the problem must be formulated. This simply means that the scientific must state exactly what the question is that he or she is attempting to answer, such as "What are the effects of failure on future tasks persistence?" or "Is there any cause-and-effect relationship between the coach's leadership style and team success?"

Second, one or more hypotheses must be posed. A hypothesis is a prediction of the answer to a previously identified question. In the first question above, the hypothesis might state that failure will cause a decrease in duration of task persistence. The hypothesis posed for the second question might be that a task-oriented coach should be more successful as compared with a socioemotionally oriented coach.

Third, an experiment is designed in which data are collected and analysed. The method of data collection varies according to the nature of the question asked. If, as in the first question above, task persistence was to be observed, then time of duration at the task would be measured.

However, if the study were designed to determine differences between leadership styles of coaches, then the data would most likely be collected through the assessment of personality by means of a standardized instrument. In sport psychology research, data are collected in a wide variety of settings, ranging from the playing field to the ski slopes to the laboratory situation. However, the analysis of those data is almost always accomplished by the use of statistical procedures. Given that the data must eventually be subjected to statistical analysis, it must be capable of taking on numerical values; it must be quantifiable. For instance, task persistence could be quantified on the basis of some unit of time such as seconds or minutes, whereas personality would take on some numerical value associated with a person's score on a particular personality scale.

Fourth, after the data are collected and analysed, the scientist draws a conclusion the basis of what the data analysis has shown. Recall that the purpose of the third step is to design an investigation that will allow the scientist derived on the basis of the data analysis tell the experimenter whether the hypothesis can be accepted or must be rejected. In our examples, the data analysis would indicate whether a failure outcome did indeed have a negative effect on subsequent task persistence and whether differences in coaching style had nay measurable effect on team outcome. The scientific method is elegant in its simplicity, yet rigorous in a scientific sense because of its inherent checks and balances.

Give the latter, and given that the scientific empiricism is the best or should be the only acceptable method of "scientifically" knowing. What could be lacking? One difficulty is that the method of empiricism has as its goal only the identification of facts. More specifically, the generates fact for their own sake, regardless of any connection or relationship to other existing facts.

In the early stages of scientific inquiry the identification of such facts is crucial. However, as more and more facts are generated, if they are allowed to remain in insolation from one another, adequate understanding of a phenomenon becomes almost impossible. Look at this problem from the point of view of a coach and consider the array of seemingly unrelated information that a coach would need to know about psychology to work effectively with just one athlete in a single-event.

The coach would need to understand why some athletes are anxious in competitive settings whereas others are not; why people with different levels of anxiety perform differently within competitive sport settings; and how an athlete's anxiety reaction can be raised or lowered. Certainly it would be inefficient to require that every coach attempt to identify and synthesis all of the known facts about each of the notions just mentioned. What is the alternative?

Variables

A variable is an entity capable of being measured, or quantified; that is, it can be assigned numerical value. In the early stages of the development of a science, much energy is put into identifying and measuring variables thought to be important to that particular domain. However, the individual variables themselves provide little information of real value; it is the interrelationships that occur between the variables that are of most concern. For instance, it is of small value to be able to say that it is possible to measure whether an athlete tends to be an extrovert or an introvert. Of more importance is the ability to specify exactly how the level of introversion or extroversion relates to successful performance on a team sport such as water polo.

Thus in the first case one variable, the degree of extroversion versus introversion, was measured. In the second case the interrelationship between the psychological variable of introversion versus extroversion and sport team performance was identified. Clearly, the second case provides more valuable information.

At its most fundamental level, research in the area of sport psychology consist of the study of two variables in an effort to establish if any relationship exists between those variables. These important relationships are called causal relationships because a change in one of the variable causes a systematic change in the other. These cause-and-effect relationships relative to individual behaviour in sport or movement settings formulate the research domain of sport psychology.

Independent variables are those thought to the cause of an observed effect. They are the variables that are usually manipulated across experimental groups in a study. For instance, suppose that you designed an investigation in which you predicted that athletes who mentally practised would perform better than similarly skilled athletes who did not. The study was then designed so that one group mentally practised basketball free throws and the other group did not. If you then measured the actual free throw performance of both groups and found that the group that had practised mentally had better free throw performance than the group that did not, you could conclude that practising mentally caused the effect of enhancing basketball free throw performance. In this example the causative, or independent variable was mental practice.

Control variables are any other variables that could effect changes in the dependent variable. In the example above, because you obviously wanted to establish a cause-and-effect relationship between mental practice and effectiveness of basketball free throw shooting, you would not want a situation in which you could not be sure that the observed effect was a result of mental practice alone. Therefore when designing your study you must eliminate or control any variable, other than the independent variable, that could effect changes in the dependent variable. In our basketball example, control variables might be variables such as initial differences in skill level or initial differences in free throw percentage between the two groups of free throw shooters.

Dependent variables are those variables that are affected systematically by a change in the independent variable. The dependent variable is measured or assessed during the collection of data. In the basketball free throw example described above, the dependent variable would be free throw performance.

2

SOCIAL PSYCHOLOGY AND SPORT

Sport psychology is a discipline that focuses upon the individual. Topics such as personality, attention, arousal, anxiety, anxiety control, cognitive intervention attribution, and motivation all apply specifically to the individual. Conversely, the sociology of sport is a discipline that focuses upon social relations, group interactions, and sport-relates social phenomena. Because groups are composed of individuals, it is often difficult to determine where psychology ends and sociology begins; hence the need for an area of study called the social psychology of sport.

1. Audience Effects in Sport

Perhaps no social-psychological effect is more important to athletic performance and outcome than the audience, or spectator, effect. The evidence is clear, for example, that there is significant advantage to playing at home in baseball, football, basketball, and ice hockey.

The perception of a home court advantage is especially evident in man's collegiate basketball and professional soccer. Many basketball conferences have adopted the policy of sending the winners of their post-season tournaments to the NCAA tournament.

In professional sport, two well-publicized examples

of the home court advantage may be cited. Sports writers have coined the phrase "Celtic Mystique," when referring to the win-loss record of the Boston Celtic basketball franchise when playing at Boston Gardens. Prior to losing game number 4 to the Los Angeles Lakers in the 1987 NBA Championship series, the Celtics had won 94 of their previous 97 games in "friendly" Boston Gardens.

The 1987 World Champion Minnesota Twins baseball team won 70 percent of its regular season home games. The Twins won all their home games when they defeated the heavily favoured Detroit Tigers for the American League Pentane and the St. Louis in the World Series. While many variable may help create the home court or home field advantage, none seem to be as important as the presence of a supportive audience.

Determining how and why an audience presence affects athletic performance is the focus of this section. Topics to be discussed included social facilitation, home court advantage, home court disadvantage and audience characteristics.

Audience Characteristics

Having determined that home advantage usually exists in sport and that this advantage is related to the presence of a supportive and interactive audience, it is time to examine characteristics of the audience. Do certain audience Characteristics lead to a greater home advantage? This question will be discussed in the following paragraphs.

Crowd Hostility

It is generally understood that a supportive and friendly crowd will help the home team. What is the effect, however, of seemingly hostile and protesting on player performance? Research by Greer demonstrated that sustained hostile spectator protest have a clearly negative impact on the visiting team. Home basketball games of two Division I basketball teams were identified and studied relative to subsequent skill performance. Following episodes of sustained fan protest, the performance of athletes was monitored for five minutes of running game time.

The result of the research showed a slight improvement in the performance of the home team, paralleled by a more significant and pronounce decline in performance of the visiting team following spectator protest.

Arise Selinger, former head coach of the women's national volleyball team, can attest to the devastating effect of a hostile audience. Arise took his heavily favoured women's team to Peru in 1982 to compete in the World Championships.

Everything went according to plan until the U.S. team played Peru. That was when the crowed took over, or officials The highly unsportsmanlike fans were armed with whistles and noise makers. Each time a U.S. player went back to serve, the noise was deafening.

"The sound came down like thunder. You could feel the vibrations. You're totally dis-oriented. It's a terrible experience," said Coach Selinger. A father of one of the athletes summed it up this way: "The team prepared for nine months to win the World Championships and they were beaten by a wireless microphone and fifteen thousand plastic whistles."

Sometimes it is not necessarily the entire crowd that is hostile, but a small and highly vocal section of the crowd. A case in points is the "Antlers," a rowdy, composed of University of Missouri-Columbia undergraduates wearing black T-shirts, goofy harts, and painted faces. The antics of the "Antlers" are at times so offensive that they have to censured by university administrators.

Crowd Size, Intimacy, and Density

There is evidence in professional baseball that crowd size makes a difference. Schewartz and Barsky, for example, demonstrated that audience size is related to performance in baseball.

The winning percentage of home teams increased as the size of the crowd increased. This effect is most pronounced when first- division home teams play visiting teams from the second division. Also based upon the Schewartz and Barsky date is the observation that sport such as basketball and ice hockey enjoy a greater home advantage than baseball and football.

Since baseball and football normally accommodate a far greater number of fans than basketball and/or ice hockey, factors such as audience density and audience intimacy may be more important than size for creating the home court advantage .

Audience density and audience intimacy are relate to how tightly packed together the fans are and how close the are to the field of play. Successful teams that opt to move out of smaller, more intimate facilities into larger once often do so at the expensive of crowed density and intimacy.

Social Facilitation

Social facilitation research is based on the notion that the presence of an audience of one or more spectators can facilitate performance. This is an appealing concept, since almost everyone has experienced the desire to perform better when friends, family, or members of the opposite gender are watching.

Research in the area of social facilitation was significantly influenced by the work of Robert Zajonc. Zajonc's classic paper on the topic remains the single most critical factor in the development of social facilitation as a field of inquiry. Based upon drive theory, Zajonc's Model proposed that the presence of an audience has the effect of increasing arousal in performing subjects.

Since increased arousal facilitates the elicitation of the dominant response, the presence of an audience will enhance the performance of a skilled individual while causing a decrement in the performance of an unskilled individual.

While Zajonc's Model generated hundreds of research investigations, it suffered a fatal from the perspective of sport psychology research. Social facilitation was defined as "consequences upon behaviour which derive from the sheer presence of other individuals". To test the model, researchers were obligated to focus on the "sheer presence" of audience with no interaction between the performer and the audience.

Since this situation rarely occurs in competitive athletics, the research general from testing the model is not easily applied to sport. Therefore, the balance of this section will be spectators is assumed. For a review of social facilitation research, the reader is referred to cox

2. Leadership in Sport

It is much easier to point to examples of great leadership that it is to explain what great leadership is. For example, in sport, it would be hard to find greater examples of leadership than such coaches as Knute Rockine, Vince Lombardi, Jihn wooden, and Pat Head Summit.

Each of these men and women had an unquenchable desire to succeed, to excel, and to win. As Lombardi put it "Winning is not everything it is the only thing!". Roger Staubach was such a great field general that his team-mates never gave up as long as he was in charge. As Billy joe DuPree said of Staubach, "He never know when it was over. At the end of a game, even if we're down by 20 points, he'll be standing there by himself trying." Jonh Wooden, the most successful coach in collage basketball history, won ten NCAA national championships.

Seven of those wins were in row, beginning in 1967 and ending in 1973. Similarly, pat Head Summit, women's basketball coach at the University of Tennessed, has enjoyed unprecedented success as a leader and coach. Coach Summit is one of the most successful coaches in American ranks with any Summit such great coaches and leaders? We may never know the precise answer to a question like this, but we can study it and try to understand the many possibilities.

In some ways, the complexity of the concept of

leadership is overwhelming. It is like puzzle that makes little sense until each piece is put in its place. In an attempt to master this puzzle, this section has been organized into three subsections.

The First addresses the major theories of leadership that have evolve. The second subsection deals with the nature of the coach-athlete interaction. The third discusses geographical location and leadership opportunity.

Geographical Location and Leadership Opportunity

Of interest to sport psychologist are situations that seem to promote leadership opportunity. For example, Grusky designed a study that hypothesised that player position in baseball was relate to leadership opportunity. Specifically, Grusky proposed that the geographical location of a baseball player was predicate of whether or not the player would later become a major league manager.

Grusky argued that the critical factor in player location was the opportunity for player interaction. Further proposed that the opportunity for high interaction was associated with centrality, communication, and the nature of the task.

The term centrality refers to the spatial location of the athletes , while communication relates to the degree of verbal and visual interaction between players. The nature of the task is closely associated with interaction, since some tasks in baseball require a high degree of catching a line drive is a highly independent task, while the infielder's task of turning a double play is highly interdependent.

In Grusky's design, the infielders and pitchers were considered to be low interactors. Looking at all major league managers from 1921 to 1941 and 1951 to 1958, Grusky determined that 97 percent of them had played major league ball. Additionally, 77 percent of the managers were from the high interaction group, while only 23 percent were from the low interaction group.

Thus playing position appeared to be highly related to leadership opportunity. Furthermore, former catchers made up 26 percent of the total sample of major league managers, but they made up only 11 percent of the original player sample during the years studied. Apparently, the position of catcher is highly related to player interaction and leadership opportunity.

Interestingly, of the six major league managers hired following the strike-shortened 1994 season, four were catchers in their playing days. Bob Boone, a manager of the Kansas City Royals, said of his playing days as a catcher, "I managed every game I played in. I just didn't have the control." Dan Euquentte, Boston Red Sox general manager, was quoted as saying,

"The catcher is involved with every pitch, every pitcher, and he's the only player on the field who has the whole field in front of him".

An important follow-up to the Grusky research was a study by Loy and Sage in which the relationship between player position in baseball and leadership qualities was again investigated. Using a sample of high school athletes, Loy and Sage hypothesized that player interaction, in terms of position, would be related to the selection of team captains and interpersonal attraction. As predicted, the results indicated that high interactors were selected as the team captains much more often than low interactor. Hihg interactors were also perceived to be better .liked and more valuable to the team than lows interactors.

Chelladurain and Carron proposed a geographical location model based on the dimensions of propinquity and task dependence. In so doing, they argued against the notion of centrality, since in many sports, being located in the geographical centre of the team cannot be deemed crucial to task dependency or interaction, For Example, the catcher in baseball and the point guard in basketball are not at the geographical centre of the team, yet they are the individuals highest in interaction potential.

Chelladurain and Carron's dimension of propinquity is associated with the observability and visibility of an athlete by team-mates. The Teacher in baseball, though not always central to the action, is definitely the most observable and visible.

The second dimension of task dependence is associated with the level of interaction required to successfully complete a task. For example, a double play in baseball is a highly dependent task requiring coordination between the short stop and the second baseman.

Using the two-dimensional model, Chelladurain and Carron reconsidered the Grusky and Loy and Sage baseball date. The baseball position highest in propinquity and task dependence is the catcher. Lowest in these two dimensions were the outfielders. Clearly, certain positions on a athletic team enjoy greater propinquity and dependence. These positions are associated with greater leadership potential and opportunity.

More recently, Fabianic reconsidered the playing position hypothesis and related it to the pausisity of monitoring managers in major league baseball. A Highly visible characteristics of professional baseball is the relative absence of minorities serving in managerial positions.

Traditionally it has been argued that blacks are seldom selected as managers because they did not occupy highly interactive positions as players.

Data from 1951 to 1980 and from 1980 alone show that the majority of baseball managers continue to be selected from the interactive positions, even thought infielders made up fewer players than non-interactive to Fabiance's analyses, managers continue to be selected from highly interactive positions. However, in 1980 minority players made up 29 percent of all players and were distributed in the high and low interactive position groups of all players and were distributed in the high and low interactive position groups in the same proportion. Based on expected representation, 9 of the total 32 managers should have been from minority groups–6 from the high interactive group and 3 from the low interactive group. Yet in 1980 there were only two managers representing minority groups. Clearly, racial bias is involved in the selection of managers in professional baseball.

While geographical location continues to be salient facto in the selection of white geographical location continues to be salient factor in the selection of white managers, this generalization does not include minorities.

Most of the geographical location research dealing with leadership and focused on baseball. However, there is no reason why the Chelladurain and Carron two-dimensional model could not be applied to football, basketball, and volleyball. Based on the research by Grusky, Loy and Sage, and Fabianic, we can predict that leadership opportunities in football, basketball, and volleyball would fall primarily to quarterbacks, point guard, and setters, respectively.

Coach-Athlete Compatibility

An important factor linked with leader effectiveness is coach-athlete compatibility, or the quality of the relationship between the coach and the athlete. Compatibility between coach and athlete has been shown to be an important determinant of team success and satisfaction. In studying coach-athlete compatibility, researchers compare behaviours of effective coach-athlete days with those of less effective dyads. Dyads may be formed and compared utilizing an instrument called the Fundamental Interpersonal Relations Orientation-Behaviour Questionnaire.

The FIRO-BQ was developed by Schutz to measure the level of affection, control, and inclusion that exists between members of a dyad. Affection refers to close personal emotional feelings between two people, while control refers to the perception of power, authority, and dominance, Inclusion refers to positive association among people and is relate to communication, openness, and two-way interaction.

Conversely, incompatible coach-athlete dyads are

characterized by good communication and the presence of rewarding behaviour flowing from coach to athlete. Conversely, incompatible coach athlete dyads are characterized by a lac a communication and rewarding behaviour. In compatible dyads, coach and athlete freely interact with each other.

There is a feeling of mutual respect, an appreciation of each other's roles, and a desire communicate honest feelings. The feelings are no present in the incompatible dyads. Rather, there is a feeling of detachment and isolation from each other. Effective and open communication cannot take place in an environment of exclusion. Compatible dyads are also characterized by coaches who consistently reward athletes for effort and performance, Rewards come in the form of praise, acknowledgement of effort, and recognition for outstanding performance. An enthusiastic "An enthusiastic "pat on the back" is characteristic of compatible coach-athlete dyads.

Several investigations have shown that significant disparity exists between coach and athlete perceptions of coaching behaviours and the environment. Coaches tend to view the environment as ideal, while athlete often see a great disparity between what is happening and what they think should be happening.

The research on coach athlete interaction suggests that the relationship between the athlete and the coach has plenty of room for improvement. In light of the lack of quality interaction between coach and athlete, Miller suggested assertiveness training for coaches to help them relate to athletes.

Assertiveness involves appropriate expression of

thoughts. Unassertive coaches, on the other hand, allow players to take advantage of them. Miller's assertiveness training module for coaches involves eight essential components and eight associated there objectives. Using a volleyball example, actual execution of the model involves the following three steps:

(i) Tell what you think should be done. "I'd appreciate it if you would master your assignment and follow the strategy now and in the future."

(ii) Describe the situation to the athlete. "Your assignment was to cover the power angle on that spike."

(iii) Tell how it affects the team. "When you follow through with your assignment, it provides the coverage necessary for an effective defence."

The three steps in Miller's training program allow the coach to maintain a quality relationship. The athlete receives specific information and is told exactly how to perform. More important, the athlete is not shamed or embarrassed by attacks on his self-esteem.

Theories of Leadership

Early interest in leadership centred on the trains or abilities of great leaders. It was believed that great leaders were born and not made. Since these early beginnings, leadership research has evolved from an interest in the behaviour of leaders to the notion of situation-specific leadership.

The notion of an evolution in leadership thought is useful, but it suggests that the early researchers were somehow naive and behind the times. But a careful analysis of the early writings of some of the great researchers reveals that they were as aware of our "modern" concerns for situation-specific leadership as we are today. For example, Metcalf and Urwick quoted Mary Parker Follett as saying, "Different situations require different kinds of knowledge, and the man possessing the knowledge demanded by a certain situation tends in the best managed business to become the leader of the moment." Stogdill expressed similar sentiment when he suggested that a successful leader in one situation may not necessarily be successful in other situations.

Perhaps the most significant contribution to understanding the various approaches to categorizing leadership theory has come from Behling and Schriesheim.

They developed a typology of leadership theory. Their typology categorizes the four major approaches to studying leadership, According to whether the theory deals with leadership traits or leadership behaviours, and whether the traits or behaviours are universal or situational in nature.

Leadership traits are relatively stable personality dispositions such as intelligence, aggressiveness, and independence. Leadership behaviours have to do with the observed behaviour of leaders and have little to do with their personalities. Traits found in all successful leaders are referred to as universal Traits, as opposed to situational traits. Situational traits and situational behaviours are those traits and behaviours that may help make a leader successful in one situation, but are of little value in another, Research reported by Cratty and Sage underscores the importance of this concept.

They conducted a study in which a fraternally

pledge class competed against a loosely organized group of students who did not know one another. The task consisted of going through a maze while blindfolded.

The priori hypothesis was that the pledge would do better because of their well-established lines of communication and leadership. In fact, the group having no previous association with each other outperformed the pledge class. The reason for this finding was due to the quality of the leadership .

The pledge group tended to rely on the pledge president, who had no specific experience with such things as navigating maze blindfold. However, from among the independent group, a leader quickly emerged who had obvious skills at navigating the maze.

The members of this group turned to him for instructions and tips that allowed them to outperform their competitors. They were not hampered by a leader who had no useful skills for this task.

(i) Universal Behaviour Theories of Leadership

Shorty after World War II the focus in leadership research turned from universal traints to universal behaviours of successful leaders. It was believed that successful leaders had certain universal behaviours. Once these universal behaviours were identified, they could be thought to potential leaders everywhere.

This approach to leadership was very optimistic, since anyone could learn to be a successful leader simply by learning certain predetermined behavioural characteristics. If these universal behaviours could be mastered, then anyone could be a successful leader. Unlike trait theory, the belief was that leaders are made, not born.

The driving force behind this approach to leadership came from two different sources at approximately the same time: Ohio State University and the University of Michigan.

Two important products or concepts emerged from the universal behaviour research conducted at Ohio State University and the University of Michigan during the 1950s and early 1960s. First was the development and refinement of the Leader Behaviour Description Questionnaire, from which most of the universal behaviour research was derived. Second was the identification of consideration and initiation structure as the two most important factors characterizing the behaviour of leaders.

Consideration refers to leader behaviour that is indicative of friendship, mutual trust, respect and warmth between he leader and subordinates. Conversely, initiating structure refers to the leader's behaviour in clearly defining the relationship between the leader and organization, Channel of communication, and methods of procedure.

These two kinds of behaviour are considered to be relatively independent but not necessarily incompatible. That is, a leader could be high in both consideration and initiating structure.

It is not necessary, according to the construct, to be high in one and low in the other. Interestingly, the Michigan studies resulted in two nearly identical universal behaviours associated with leadership.

The two general dimension of leadership behaviour

identified in the Ohio State and Michigan studies have provided a basic framework for many leadership studies. Often the terms initiating structure and consideration have not been use, but compatible terms have been.

Yet the general nature of these two categories has resulted in confusion about the terms used to describe them. Leadership styles with such labels as democratic, egalitarian and employee oriented reflect nature tend toward behaviour that can best be explained in terms of initiating structure or production emphasis. Leadership styles that are basically democratic in nature tend toward behaviour that can best be explained in terms of consideration and employee orientation.

(ii) Fielder's Contingency Theory

Fielder's contingency theory provides an excellent example of a leadership theory that is situation-specific, but retains the notion of personality traits. Fielder's theory is one of many that uses the contingency approach.

The contingency approach to leadership suggests that leader effectiveness is somehow situation-specific, and that leader behaviours that ate effective in one situation may not be in another. In a sense, effective leadership depends on specific environmental situations. However, Fiedler's theory differs from most situational theories, since the emphasis is on relatively stable personality traits, as opposed to behaviours. Thus, a particular personality disposition that seems to be effective in one leadership situation may not be effective in another.

According to Fiedler, the contingency model of

effective leadership posits that the effectiveness of a group is contingent on the relationship between leadership style and the degree to which the situation enables the leader to exert influence.

The theory holds that the effectiveness of a group depends on two factors: the personality of the leader, and the degree to which the situation gives the leader power, control, and influence over the situation.

In terms of the personality, Fiedler believes that leaders are either relationship motivated or task motivated. Relationship between leader and followers. Successful performance of the task is of secondary importance to this type of leader. Task motivation, on the other hand.

The satisfactory completion of the task is important to this type of leader, while establishing and maintaining positive interpersonal relationships is secondary.

To measure these two personality types, Fiedler developed the Least Preferred scale, which measures the leader's empathy for her least preferred team member. A high score on the LPC would indicate that the leader is able to have positive feeling towards a weak or nonproductive member of the group, and thus is relationship motivated. A low score on the LPC would indicate that the leader is unable to rate the least preferred co-worker very high, and thus is task motivated.

The second major factor in Fiedler's contingency model is situational favourableness. This construct indicates the degree to which the situation gives the leader control and influence over the environment. According to Fielder, situational favourableness depends upon three sub-factors: leader-member relations, task structure, and leader position power.

The horizontal axis indicates the eight cells of the situational favourableness dimension, with the left side of the graph representing the most favourable situation and the right side the least favourable situation.

The vertical axis represents performance effectiveness for the relationship-oriented person and the task-oriented person. The two curves represent fiedler's basic predictions. Specifically, relationship-oriented leaders perform best in situations of moderate favourableness, while task-oriented leaders perform best in either favourable or unfavourable situations.

Fielder's contingency theory has intuitive appeal and some research support. Perhaps the most controversial aspect of the theory is the basic proposition that leadership training programs are of little value. Leadership training programmes only help the leader learn how to enhance power and influence. However, increased power and influence would not benefit the relationship-oriented person, who does best in a moderately favourable situation. Therefore, Iiedler proposes that there are only two ways to improve leadership effectiveness.

The first involves changing a leader's personality. This is unlikely to happen, since core personality dispositions cannot be easily changed. The second approach involves modifying the degree to which the situation is favourable to a certain type of leader. Fiedler suggests that this could be done by adjusting some aspect of organizational structure, or looking for leaders who possess personality characteristics consistent with existing structure and situational favourableness.

Intuitively, there are many example in which fielder's theory could apply in sport. Doug Colling, a former coach of the World Champion Chicago Bulls basketball team, provides a good example. Because of the nature of his emotional volatility, he was able to take a young team to near stardom during his early years with the Bulls.

His volatile personality was useful in motivating a young and unpredictable team. However, this same personality characteristic became a liability as the Bulls matured as a team and began to tune him out. History records that Collins was fired after the 1988-89 season and replaced by Phi Jackson.

A Second example would be Johny Keane. Following the failure of the New York Yankees to win the World Series in 1964.

The Yankees finished the 1965 season with a losing recorded of 77-85, and Johny Keane was fired the following year.

(iii) Situation-Specific Theories of Leadership

Many contingency theories of leadership, or theories that hypothesize an interaction between the leader and the situation, have been studied. The basic difference between Fiedler's contingency theory and those that are to be discussed in this section is that Fiedler insisted on looking at relatively stable personality traits, as opposed to behaviours.

The theories in this section view leadership as a

function of the interaction between leader behaviour in a specific situation and the situation itself.

Some of the situation-specific theories are path-goal theory ; adaptive-reactive theory; the role making. Unfortunately, space does no allow a thorough review of each of these theories; the first two will be considered because of their potential application to athletics.

In addition, Chelladural's multidimensional model leadership, and Smoll and Smith's model of Leadership behaviours in Sport will be introduced as examples of leadership theory that has evolved from the sport science.

Path-goal Theory In Fiedler's theory, the emphasis was on the personality of the leader and the favourableness of the situation. In Path-goal theory, the emphasis is one the needs and goals of the subordinate or the athlete. In other words, the leader is viewed as facilitator.

The coach or leader helps athletes realize their goals. The leader's success is viewed in terms of whether or not the subordinate achieve their goals. Thus, the basic proposition of path-goals theory is that the function of the leader is to provide a "well-lighted path" to assist the follower in achieving goals.

This is done by rewarding subordinates for goal attainment, pointing out roadblocks and pitfalls on the path to success, and increasing the opportunities for personal satisfaction. For example, if an athlete's goal is to break a school record in the mile run, it is the coach's jot to provide a training program that is rewarding and enables the athlete to accomplish this goals.

Path-goal theory has not been investigated much either in or out of sport environments, perhaps due to its lack of clarity. However, Chelladurian and Salech Looked at the theory from a sport context and reported partial support for path-goal theory.

Individuals who demonstrated a preference for team sports also indicated preference for leader behaviour that was calculated to improve performance through training procedures. Thus, leader behaviour correlated with the athlete's particular athlete personality consistently preferred a particular leader behaviour.

Life Cycle Theory Like path-goal theory, life cycle theory places the emphasis in leadership behaviour on the subordinates and not on the leader. The appropriate leadership style for any specific situation depends on the maturity of the subordinate. Two Types of leadership behaviour, conceptualized in terms of relationship behaviour and task behaviour, are possible.

The appropriate combination of task and relationship behaviour depends on the maturity of the follower. Maturity is defined in terms of the capacity to set and obtain goals, willingness and ability to assume responsibility, and education and/or experience. According to this model, the need for task structure behaviour decreases with increased maturity. However, the need for relation ship behaviour forms an inverted U relative to maturity. At low and high levels of maturity, relationship behaviour should be low, but at the moderate levels of maturity it should be high.

Case reported an investigation in which the basic tenets of Hersey and Blanchard's life cycle theory were tested in an applied setting. Leadership behaviours of successful head basketball players completed the LBDQ regarding their respective coaches.

Athletes were categorized according to the maturity dimensions of junior high, high school, college, and A.A.U. participation. Initiating structure and consideration leadership scores were calculate for 40 coaches and then categorized according to maturity level.

The results revealed a quadratic relationship between maturity and relationship behaviour and between maturity and task behaviour. Task behaviour of successful coaches was low at the junior high and A.A.U. levels of maturity and high for the high school and collegiate levels. Conversely, relationship behaviour of successful coaches was high school and collegiate levels.

The Case model requires additional testing before it can be used for prediction purposes. Intuitively, the model is appealing because it is somewhat consistent with expectations. Yet, these same athletes might be expected to respond well to an environment in which concern for the athlete was openly expressed. Moderately mature athletes would require a fair amount of task structure in order to perform well. In this case, relationship behaviour might appear to be less pronounced and less important.

Chelladurain's Multidimensional Model of Leadership Chelladurain's multidimensional model of leadership. In this model, athlete satisfaction and performance are viewed as the products of the interaction of three components of leadership: prescribed leader behaviour, preferred leader behaviour, and actual leader behaviour, Prescribed leader behaviours and those that conform to the established norms of the organization.

In the military, for example, officers are expected to behave in a certain manner in the presence of their subordinate. Preferred leader behaviours are those behaviours that are preferred by the coach socialize with the team after a game. Finally, actual leader behaviours are those behaviours that the leader exhibits, irrespective of the norms or preferences of the team.

General support for Chelladurain's model has been forthcoming. Athletes associated with specific sports and cultures tend to prefer a certain type of leadership style When the coaches of these sport teams exhibit leadership styles that are incongruent with an athlete's preferred leadership style, athlete performance and/or satisfaction may decline.

The type of leadership style preferred by athletes varies as a team cohesion perceive their coach as possessing a number of desirable coaching characteristics. Leadership Behaviour Model As proposed by Smoll and Smith, the Leadership Behaviour Model is based upon situation-specific behaviours of the leader. The dotted lines in the model represent the effect of various mediating variables upon the central process. In the model, coach individual difference variables include such factors as goals, intentions, perceptions of self-athletes, and gender. Player individual difference variables include such thing as age, gender, perceptions about coach, motivation, anxiety, and self-confidence. Situational factors include things like nature of the sport, competitive level, success/failure, and team cohesion.

The dotted line leading from player individual difference variables to coach behaviours was added to the model to reflect recent research by Kenow and Williams. Coach behaviour is influenced by the coach's perception of the individual athlete. A coach may treat an athlete who exhibits high anxiety or low self-confidence differently from other athletes. The model provides a Framework for leadership behaviour research in sport.

The science of observing and recording coaching behaviour has been refined through the development of various standardized scoring systems. The most thoroughly studied systems is the Coaching Behaviour Assessment System, developed by Smith, Smoll, and Hunt.

The observed behaviours are reactive and spontaneous in nature. Reactive behaviours are coach reactions to player or team behaviours. For example, a layer makes a mistake and the coach responds by verbally chastising the player. Spontaneous behaviours are initiated by the coach and do not occur in response to a player behaviour.

Research with the CBAS has revealed number of interesting relationships. When they are working with youth sport athletes, the dominant behaviours of coaches are positive reinforcement, general technical instructions, and general encouragement.

The behaviours of keeping control and administering punishment are perceived by players to occur much more often than they actually do. Another interesting finding is that coaches of youth sport teams spend a greater amount of their time providing technical instruction and feedback to low-expectation youth than to high expectations youth. In other words, the coach does not favour the athletes the expect to be the better performers.

The CBAS has also been utilized as a measurement tool to determine if behavioural training programs are effective in teaching youth sports coaches to be better and more effective leaders of youth. Research has demonstrated that well conceived, well-planned efforts to train coaches are effective. Desirable coaching behaviours can be identified and conveyed to new coaches.

Furthermore, improve coaching behaviours of coaches result in benefit to the young athletes in the form of greater satisfaction and reduced anxiety.

While the CBAS has been the most widely studied system for observing and documenting coaching behaviours in youth sports, it is not the only one. A case in point is the Arizona State University Observation Instrument developed by Lacy and Darst. The ASUOI is compose of 17 behavioural categories 7 of which are directly related to instruction.

It has been utilized in studying coaching behaviour in American football, tennis, gymnastics, basketball, and soccer. Research with the ASUOI has tended to focus upon behaviours of coaches of adult athletes as opposed to those of youth sport athletes.

(iv) Trait Theories of Leadership

Trait theory has its origin in the "great man" theory

of leadership, which suggests that certain great leaders have personality traits and personality characteristics that make them ideally suited for leadership. The heyday of trait leadership theory began with the development of objective personality tests in the 1920s and lasted until the end of world War II. Proponents of trait theory believe that successful leaders have certain personality characteristics or leadership traits that make it possible for them to be successful leaders in any situation.

Since these personality traits are relatively stable, it should be possible to identify potential leaders simply be administering a personality inventory. This approach had a great deal of support from social scientists prior to and during World War II, but after the war, support waned rapidly.

The beginning of the decline in treat leadership theory occurred shortly after Stogdill published his review of 124 trait-related studies. His review and general conclusions led social scientists to discredit the universal trait theory of leadership.

It was simply not possible to demonstrate that successful leaders possessed a universal set of leadership traits. A comparable review of sport-related literature led Sage to make the same conclusion relative to leadership in sport.

Since the early 1970s there has been a sharp decline in the number of sport studies investigating trait leadership theory. In some ways this is appropriate, since there does not seem to be universal set of personality traits that would set the successful leader or coach apart from less successful colleagues. However, in some respects this is also unfortunate. While there may not be a universal set of traits associated with successful leadership, this is not to say that certain combinations of traits might not be beneficial in specific situations. Such an approach is evident in Fiedler's contingency model and in some of the situational theories.

In this theories, personality traits and/or characteristics behaviours are studied in light of the situation. Personality traits and behaviours that lead to successful leadership in one situation are observed to be of little value in another.

3. Aggression in Sport

At New York's Shea Stadium during a 1978 Jetsl-Steelers football game, spectators overpowered a security guard and dropped him over a railing to concrete walkway 15 feet below. In Toronto, four members of the Philadelphia Flyers hockey team faced maximum penalties of up to three year in prison on assault charge stemming from a wild brawl during the Stanley Cup play-offs of 1975. In the same year, Henry Boucha suffered a severe beating from Dave Forves during a Boston-Minnesota hockey game, causing boucha to lose 70 percent of his sight in one eye.

In professional basketball, on December 9, 1977, Los Angeles Laker Kermit Washington literally shattered the fact of toward Rudy Tomjanovich of the Houston Rockers with a devastating punch. On August 13, 1978, football fans were shocked by the crushing blow Jack Tatum game receiver Darryl Stingley in an exhibition match between two NHL professional hockey teams.

The Calagary Flames took a brawling home-and-home series from archival Edmonton Oilers. The Flames won the first game 5 to 4 and the second 6 to 3. While the first game was a reasonable facsimile of fair play, the second was nothing but a brawl. In 191, Rob Dibble, pitcher for the Cincinnati Reds major league baseball team, aimed his fast-ball at base runner doug Dascenzo as he ran out a successful bunt. During a 1995 Houstaon Rockets professional basketball game against the Portland Trailblazers, the Rockets' Vernon Maxwell went into the stand in Portland and slagged an abusive fan. Maxwell was suspended for ten days and fined $ 20,000.

Fighting and violence among athletes in not limited to men. After Missouri's 72-70 home court basketball victory over rival Oklahoma on January 17, 1987, a brawl erupted. The centre of attention was Oklahoma's volatile coach. With players slagging and kicking, the coach was observed on her back kicking an opposing player.

Most of the examples of aggression and violence described involved participating athletes. However, numerous examples of violence can be cited that involved spectator. One of the most repugnant examples of fan violence occurred in 1985 in Europe, where a soccer riot in Brussels left 38 dead and 437 injured after English hooligans attacked panic-stricken Italian Fans.

The riot occurred prior to the European Cup soccer final in Heysel Stadium in Brussels between Liverpool an Juventus, the soccer team of Turin, Italy. Well-liquored Liverpool hooligans attacked Juventus fans with broken bottles, tin cans, flag stick, and metal bars. Within minutes hundreds of Italian fans found themselves pressed against a chain-link fence and a restraining wall. As more bodies pressed against the barriers, they collapsed, pitching hundreds of terrified fans in which 31 Italians, 4 Belgians and 2 Frenchmen, were killed, most by suffocation.

The event has since been referred to as Black Wednesday by showed residents of Liverpool Similarly, more than 40 people were killed and 50 injured at an exhibition soccer match in Johannesburg. South African in 1991. Most of the deaths occurred when panicked spectators were crushed against a fence around the field and trampled by fleeing people. Two children were among the dead. Most recently, following a 1996 heavyweight boxing match in New York between Riddick Bowe and Andrew Golata, a confirmed riot ensued.

The fight ended after the seventh round as Golata was disqualified for throwing his fourth low blow to Bowe's groin. Thirty-five minutes after the bout, the crowd was ordered to evacuate Madison Square Gardends as riot police rushed in.

A number of critical questions come to mind as one contemplates the issues of sport aggression. Does participating in or observing violent sporting events merely teach and encourage further aggression on and off the playing field? If these two questions can be answered, then is it possible to eliminate aggression and violence from sports? If so, how?

In this section these questions and other critically important issues will be discussed. Topics to be addressed include defining aggression, theories of aggression, the catharsis hypothesis, measurement issues, fan violence, performance issues, situational factors contribution to aggression, and reducing aggression in sport.

Defining Aggression

Two factors must be present in order for a behaviour to be labelled aggression. First, the behaviour must be aimed at another human being with the goal of inflecting physical harm. Second, there must be a reasonable expectation that the attempt to inflict bodily harm will be successful. Consequently, the following behaviours, often mislabelled aggression, are not really examples of aggression:

(i) Aggressive behaviour in which there is no chance for the intended victim to be injured.

(ii) Doing destructive violence to an inanimate object such as a door or a water cooler.

(iii) Un-intentionally injuring another person during athletic competition.

Over the years, two basic kinds of aggression have been identified. The first is hostile aggression. For individual engaged in hostile aggression, the primary goal is the injury of another human being. The intent is to make the victim suffer, and the reinforcement is the pain and suffering that is caused.

This sort of aggression is always accompanied by anger on the part of the aggressor. A good example of hostile aggression occurs when a baseball pitcher throws a high-inside fast-ball at a batter who has angered him.

The clear attempt to injure is present, and the goal is to cause suffering. The outcome of the contest is not factor to be considered. The goal is to harm, not to win.

The second major kind of aggression is instrumental aggression. For individuals engaged in instrumental aggression, the intent to harm another individual is present, but the goal to realize some external goal such as money, victory or prestige. The aggressor view the aggressive act as instrumental in obtaining the primary goal. A parallel baseball example for instrumental aggression would be one in which the pitcher has been "ordered" by his manager to hit a batter in retaliation for some earlier infraction. The pitcher is not necessarily angry at the batter, but sees hitting the batter as instrumental in achieving the team goal of winning the game.

It must be emphasized that neither type of aggression acceptable. The aggressor is guilty of purposely inflicting harm with the intent to injure another person. This must be discouraged at all levels of competition, especially the professional level, because young athletes everywhere emulate the pros.

A Third category of behaviour that is often confused with aggression is assertiveness, or assertive behaviour. Generally, when coaches encourage their athletes to be more aggressive, what they really want is that they be more assertive. Coaches want their athletes to assert themselves and make their presence felt. Assertiveness involves the use of legitimate physical or verbal force to achieve one's purpose. However, there is no intent to harm the opponent. Even if an opponent is harmed as a result of a tackle in soccer, it is not necessarily aggression. It is merely assertive play, as long as it is within the spirit of the agreed-on rules and the intent to harm is not present.

Assertiveness requires the expenditure of usual effort and energy, but if there is no intent to harm, then any resultant harm is incidental to the game. Let's go back to our baseball example. If the pitcher throws a high inside fast-ball, with no intent to hit the batter, this is considered assertive play. The pitcher must establish control of the strike zone, or the batter will intimidate the pitcher into throwing either strikes or outside balls.

This is to be expected, since at times only the athlete knows whether an "aggressive" act was intended. From a practical standpoint, it is the job of the official to penalize any behaviour that is in violation of the rules, regardless of the intent of the violator.

However, most sports make provisions for extraordinary penalties if the behaviour is deemed to be intentional and/or dangerous. For clarity, let's return to the baseball pitching example. If in the judgment of the umpire a pitcher purposely throws a pitch at a batter with intent to harm, he must be penalized, regarded of why he did it. However, if a batter has body over the strike zone, he is inviting an assertive pitcher to throw a fast-ball over the inside part of the plate. If the batter is hit, it is not the fault of the pitcher and it not an example of sport aggression.

The Catharsis Effect

Introduced earlier, the catharsis effect represents a release of pent-up frustration that makes one feel better. It is a purging of the anger and frustration associated with not being able to accomplish a goal. Venting your frustration upon punching bag or some other suitable object many provide an excellent release of anger and pent-up emotion. Venting your frustration upon a punching bag or some other suitable object may provide an excellent release of anger and pent-up emotion.

Venting your frustration upon another human being may also represent a catharsis or venting of pent-8-negative affect. In this case, however, the behaviour is considered to be inappropriate and counter to codes of proper social conduct.

If an act of aggression is associate with personal gratification there is a good chance that it will occur again. This is one way in which aggression can become a learned behaviour. Evidence from social learning theory suggest that engaging in or observing acts of aggression results in learned aggressive behaviour.

Situational Factors in a Sport Setting

Much of the research in sport-related aggression has dealt with situation-situation-specific factors. Factors associated with the occurrence of aggression in sport-specific situations are as follow:

(i) Structure of the game. Two of the earliest studies of game variables and aggression were conducted by Volkamer with soccer and Wankel with ice hockey. These studies were quickly followed by investigations by Lefebvre and Paser with soccer, Cullen and Cullne with ice hockey, Martin with basketball and wrestling, Russell and Drewery with soccer. Following is a summary of the findings with respect to game variables:

(a) Outcome of participation. Consistent with the frustration-aggression hypothesis, members of losing teams are observed to be more aggressive than members of winning teams.

(b) Point differential. More aggressive penalties occur as the game score differential increase. When teams are tied or the scores are close, aggression is at a minimum.

The penalty for aggression in a critical game situation is so high that players, coaches, and managers go out of their ways to avoid it.

(c) League Standing. The lower a team is in the standings, the more its members engage in aggression. The lowest incidence of aggression occurs with teams in first place.

(d) Playing at home or away. Whether home or visiting teams display more aggressive behaviour may depend on the nature of the aggression and the type of game involved. Soccer teams tend to be more aggressive when playing away from home, whereas aggression is almost equal for home and visiting ice hockey teams.

(ii) Environmental temperature. Using archival date from major leaguer baseball games players during the 1986, 1987 and 1988 seasons, Reifman, Larrick, and Fein observe a linear relationship between hit batters and environmental temperature. The data suggest that higher temperatures lead major league pitchers to become move aggressive in pitching to batters.

(iii) Fear of retaliation. To some degree, the fear of retaliation on the part of the individual who is the target of aggression can inhibit another player from initiating that aggression. A basketball player is a little less likely to elbow her opponent in the fibs if she fears similar treatment from the opponent.

This sort of respect for an opponent's ability to "give as good as she gets," however, can quickly escalate into open aggression and counter-aggression.

Theories of Aggression

A number of theories have been proposed to explain

the phenomenon of aggression. These theories fall into four main categories: instinct theory, social learning theory, Bredemeir's theory of moral reasoning, and Berkowitz's reformulation of the frustration-aggression hypothesis.

(i) Reformulated Frustration-Aggression Theory

As originally presented by Dollard, Miller, Doob, Mouredr, and Sears, Frustration-aggression theory proposes that aggression is a natural response to frustration and that the aggressive act provides a catharsis, or purging, of the anger associated with the frustration. Berkowitz's reformulation of frustration does not necessarily result in aggression and proposed that frustration creates a readiness for aggression. For aggression to actually occur, certain stimuli associated with aggression must be the "re flag" for the enraged and frustrated bull. In the presence of frustration, certain stimuli can serve as "triggers" to release a disposition towards aggression in a frustrated individual. Negative affect, associated with frustration, is the fundamental spur to the inclination for aggression. Anger is the root of hostile aggression, but depression is another example of negative affect that can trigger an aggressive act.

The development of aggressive tendencies is complex, but certainly learning from parents, and other aggressors is a paramount

(ii) Bredemeir's Theory of Moral Reasoning and Aggression

Based upon jean Piaget's theory of cognitive development, Bredemeir's theory of moral reasoning proposes that an individual's willingness to engage in aggression is related to her stage of moral reasoning.

Since human aggression is viewed as unethical. Bredemeier reasoned that a relationship should exist between level of moral reasoning and overt acts of athletic aggression. Contact sport, because it legitimizes acts of aggression, may actually retard a person's moral development.

This level of morality necessary for everyday life is often suspended during athletic competition,. Breadeneier refers to this suspension of ethical morality as braceted morality. Furthermore, athletic teams create a "moral atmosphere" that may be conducive to the willingness to agrees.

(iii) Social Learning Theory

Social learning theory posits that aggression is a function of learning, and that biological drive and frustration are inadequate explanations of the phenomenon. While the notion of a catharsis is an important competent of both biological instinct theory and frustration-aggression Theory, it has no place in social learning theory. Acts of aggression serve only to lay the foundation for more aggression and do not result in a reduction or purging of the drive to be aggressive. Perhaps the leading advocate of social learning theory, relative to aggression, is Bandura. Bandura has argued that aggression has a circular effect.

That is, one act of aggression leads to further aggression. This pattern will continue until the circle is broken by some type of positive or negative reinforcement. Smith, for instance, argues that violence in ice hockey is due to modelling. Youngsters learn aggression by watching their role models, the professionals, on television or in person. As long as aggression in professional sports is tolerated, children will continue of have adult models of aggressive behaviour.

(iv) Instinct Theory

Instinct theory is based upon the writings of Sigmund Freud and ethologists such as Konrad Lorenz. Freud viewed aggression as an inborn drive similar to hunger, thirst, and sexual desire. According to Freud, aggression is unavoidable since it is innate, but as with any drive it can be regulated through discharge, or fulfilment.

Since humankind is innately aggressive, it benefits society to promote athletic sports and games that provide a socially acceptable for aggression. An important corollary of the biological instinct theory is the notion that aggression results in a purging, or releasing, or the aggression drive.

This burping of pent-up aggression is known as catharsis. According to instinct theory, striking an opposing player serves as a catharsis or release of pent-up aggression.

Reducing Aggression in Sport

Aggression in sport can be curtailed, or at least minimized, if all concerned are interested in doing so. The sad part is that some of the most influential people actually promote rather than discourage violence because they believe it sells tickets. As this attitude is allowed to continue, there is little hope of solving the problem. If it is allowed to continue on the professional level, it will continue to be promoted at the lower skill levels, Athletes in the youth leagues emulate their heroes on the collegiate and professional levels.

They watch their sport heroes receive awards, applause, money, and adulation for behaviour that borders on open aggression, and they want to become like them. In this section we will discuss ways to reduce aggression on the part of participants and spectators.

(i) Curtailing aggression by spectators. The following practices would help alleviate spectator aggression at sporting events:

(a) The media should not promote or encourage the perception of friction or hatred between two teams prior to competition.

(b) The sale, distribution, and use of alcoholic beverages at sporting events should be limited.

(c) Athletic events should be promoted as family affairs.

(d) Responsible media can discourage aggression in sport not glamorizing it.

(e) Interaction between members of opposing teams should be encouraged by coaches and managers during the days leading up to a contest.

(f) As with the athletes and coaches, spectator aggression must be swiftly and severely punished.

(ii) Curtailing aggression by athletes. In this section, strategies that can be implemented to curtail athletes aggression will be identified. Based on the literature, the following recommendations can be made:

(a) Young athletes must be provide with models of non-aggressive behaviour.

(b) Strategies and coping skills designed to curtail acts of aggression should be practised.

(c) External stimuli or cues capable of evoking hostile aggression on the field of play should be removed.

(d) Coaches who encourage or even allow their athletes to engage in aggressive acts should be fined, censored. And/or suspended from their coaching duties.

(e) An athlete who engages in an aggressive act must be severely penalized.

(f) Coaches and referees should be encouraged to attend in-service work shops for dealing with aggression and violence on the part of players.

(g) Along with punishment for acts of aggression, athletes must also receive positive reinforcement for controlling their tempers in highly volatile situation.

Effects of Aggression on Performance

Conventional wisdom argues that acts of aggression on the part of an athlete will constitute a distraction and result in a decrement in performance. Not only are aggressive acts on the part of an individual distraction to the individual, but they are likely to be distracting to the team as a whole, Research shows, for example, that the lower a team is in the standings, the more likely it will be to engage in aggression. An exception to this general observation may be in the sport of ice hockey, where it is often difficult to separate legitimate force from illegitimate force.

Because hostile aggression is associated with anger and hence with an increase in physiological arousal, it is possible that for some individuals an increase in anger may produce a level of physiological arousal conducive to best performance. Conversely it may produce a level of arousal that is above an athlete's zone of optimal functioning.

Fan Violence

Most of our discussion in this section has focused upon acts of aggression on the part of sport participants. Unfortunately, however, some of the worst examples of sports aggression and violence occur among the fans watching an athletic contest.

Intense rivalries, nationalism, and alcohol abuse are major factors contributing to fan violence. Every sports event is attended by individuals who may instigate fan violence. These are individuals who score high in the personality dispositions of anger and physical aggression. These individuals are attracted to violence and fighting among fans, and exhibit a false belief about the willingness of other fans to join in acts of violence.

The false consensus effect emboldens individuals with a disposition for violence, to believe that other fans share their infatuation for fighting and would willingly join them in precipitating an altercation.

Measurement of Aggression

A number of inventories have been developed to measure aggressiveness as a personality disposition or trait. The two most current inventories include the Aggression Questionnaire and the Aggression Inventory. Sport-specific inventories include the athletic Aggression Inventory and the continuum of Injurious Acts

The actual measurement of aggression, however, is

much more difficult. Because aggression is defined as the intent to harm another human being, the measurement device must be able to capture this intent. This approach generally requires the use of trained observers using a standardized checklist of some kind.

4. Team Cohesion

Albert Carron, a prominent sport social psychologist, defined group cohesion as "dynamic process which is reflected in the tendency for a group to stick together and remain united in the pursuit of goals and objectives". Because an athletic team is a group, Carron's definition of group cohesion applies equally well as a definition for team cohesion, Intuitively, team cohesion is the elusive ingredient that changes a disorganized collection of individuals into a team.

Fundamental to the study of team cohesion is the understanding of group dynamics. Members of a team or group begin to interact with each other the moment the group is first formed. Once a group is formed, it ceases to interact with outside forces in the same manner that a collection of individuals would. The team becomes an entity in and of itself. From a Gestalt perspective, the whole is greater than the sum of its parts.

In sport, it is well-established principle that a group of individuals working together is far more effective than the same individuals working independently of one another. On a basketball team, there may be several individuals capable of scoring 20 or more points a game, However, in the interest of team success, the coach may require that one or more of these athletes assume non-scoring roles. For example, a point guard has the primary responsibility of setting up plays and getting the offence started, while the power forward must "crash" the boards and get offensive and defensive rebounds. Athletes who played these specialized roles rarely score as many point as shooting guards or forwards.

Yet, our of the desire to be ""team players," these athletes accept less glamorous roles for the common good of the team. Thus, as a group or team evolves, a certain structure develops.

This structure varies from group to group `and situation to situation, but it is critical for team success.

Not only do members of successful team have the ability to work together; they also enjoy a certain attraction to one another. In this respect, it seems logical that teams composed of members who like each other and enjoy playing together will somehow be more successful than teams lacking this quality. In 1979, the Pittsburgh Pirates won the world Series. Their theme was ""We Are Family," suggesting that they owed their success to this ability to get along and work together for a common goal. Ironically, the Auckland Athletics of the early 1970s and the New York Yankees in 1978 also enjoyed World series success, but with well-publicized disharmony within their ranks.

The team attraction necessary to bring the best out of a group of athletes appeared to be lacking in the 1988 cincinnati reds major league baseball team. After being picked to win the National league West, the Reds, led by Pete Rose, suffered a lacklustre season. In attempting to explain how the talent, laden team could perform so poorly.

As a social psychological topic, team cohesion ranks

as a very important factor for enhancing team performance and feelings of satisfaction among members. It has evolved as a fairly complex concept that requires study and additional research before it can be thoroughly understood. In the following paragraphs, team cohesion will be discussed as a function of measurement, determinants, consequences, and development.

Developing Team Cohesion

A sport team can be viewed as having both social, or interpersonal, group activities and group task-related activities. As teams develop social and task cohesion, they proceed through four basic stages. These stages are referred to by Tuckman as forming, storming, norming, and performing. Thus, it is only in the final stage of team cohesion development that it would be possible to discriminate between two groups in terms of task performance. If teams have not worked together long enough to develop team cohesion, it is unlikely that cohesion would be a factor in helping a team accomplish its goals,. During the forming and storming stages, team cohesion during the two stages would tend to be unreliable and unstable.

During the norming and performing stages, team members begin to established relationships and develop an awareness of one another's strengths and weaknesses. It is during these two latter stages that the measurement of team cohesion will be most meaningful.

Therefore, the practice of measuring team cohesion early in the season or before the season starts might not result in reliable measure of cohesion. It would be better to wait until mid-season, after team members have passed through the forming and storming stages of developing team cohesion. However, if the stability of the team remains basically unchanged from season to season, this may not be serious problem.

Professional sports teams tend to remain relatively stable across one or two seasons, whereas high school teams might change significantly from one years to the next.

High levels of team cohesion are associated with a host of positive individual and group consequences. For this reason, it is desirable that steps be taken to develop cohesion in sport teams.

This can be accomplished through team building. Team building is a process by which the coach or leader develops and carries out strategies designed to increase social and team cohesion among members. The following general principles are suggested for developing team cohesion in sport teams.

(i) Develop a feeling of "ownership" among the players, Individual players need to feel that team is their team and not the coach's team. This is accomplished by helping players become involved in decisions that affect the team and them personally. Individual players need to feel that their voices will be heard.

(ii) Acquaint each player with the responsibilities of other players. This can be accomplished by allowing players to play other positions during practices. This will give them an appreciation for the importance of other team players. For example, a spikier in volleyball who complains of poor setting should be given the chance to set once in a while.

(iii) Set team goals and take pride in their accomplishments. Individuals and teams as a whole must have a sense of direction. Challenging but obtainable goals should be set throughout the season. When these goals are reached, players should collectively be encouraged to take pride in their accomplishments and then set more goals.

(iv) Highlight areas of team success, even when the team loses a game or match. Since we know from the literature that performance affects feelings of satisfaction and cohesion, the coach must capitalize on this. If a volleyball team played good team defence in a losing effort, point this out to them.

(v) Since cliques characteristically work in opposition to the task goals of a team, their formation must be avoided. Cliques often form as a result of (a) Constant losing, (b) players' needs not being met, (c) players not getting adequate opportunities to play, and (c) coaches who promote the development of cliques through the use of "scapegoats" or personal prejudice.

(vi) As a coach or teacher, take the time to learn something personal about each athlete on the team. People will come to appreciate and cooperate with those who know little things about them, such as a girl friend's name, a birthday, or a special hobby.

(vii) Develop pride within the sub-units of large teams. For example, in football the various special teams need to feel important to the team and take pride in their accomplishments. For smaller units such as basketball teams, this may not be as critical. However, the team as a whole should develop pride in its accomplishments.

(viii) Set team goals and take pride in their accomplishments. Individuals and teams as a whole must have a sense of direction. Challenging but obtainable goals should be set throughout the season. When these goals are reached, players

(ix) Do not demand or even expect complete social tranquillity. While it is not conducive to team cohesion to allow interpersonal conflicts to disrupt team unity, it is equally unrealistic to expect interpersonal conflicts to disrupt team unity, it is equally unrealistic to expect interpersonal conflicts to be completely absent. Any time individuals are brought together in a group, there is potential for conflict. The complete elimination of any friction may actually suggest a complete lack of interest in group goals.

(x) Develop team drills and lead-up games that encourage member cooperation. Many drills are designed solely for the purpose of skills development. Many other drills must be developed that teach athletes the importance of reliance upon team-mates. For example, in basketball, drills that emphasize the importance of team-mate assists could be emphasized.

Measurement of Team Cohesion

Early attempts to measure cohesion led to confusion about the nature of team cohesion, as well as its relationship with other variables. The confusion was due primarily to two factors.

This first had to do with the failure to discriminate between task and social cohesion, and the second, with the failure to discriminate between attraction to individual members of the group goals and experience success in obtaining these goals. Groups that are high in social cohesion tend to consider social interaction and member attraction to be more important than group goals.

A measurement inventory that fails to discriminate between these two dimensions of team cohesion cannot accurately represent its multidimensional nature. The result of an investigation in which task and social cohesion were measured independently could differ markedly from an investigation in which the two aspects of cohesion were confounded.

It is quite easy to see how these two types of team cohesion could operate independently. Consider the world champion New York Yankees of 1978. This team was the best in the world in task cohesion.

They could turn double plays, hit cut offend, advance runners, and work together better than any other baseball team in the world. Yet their social cohesion was very low. Team members fought with each other, cliques were formed, and angry words were exchanged.

A measurement inventory that fails to differentiate between individual attraction to the group and group integration could be hopelessly confounded. Studies using indirect measure of team cohesion have resulted in conclusions different from those of studies that have measured team cohesion directly. Team cohesion is measured indirectly when members are asked about their assessment of the group as a whole.

The measurement problems associated with team cohesion have been addressed by the development of a conceptual framework from which a multidimensional team cohesion inventory could be fashioned

The conceptual framework, illustrated resulted in the development of the 18 item Group Environment Questionnaire. The GEQ Measures four dimensions of team cohesion: Group Integration–Task ; Group Integration–social. The following is a sample Group Integration–Task question from the questionnaire:

Since the development of the GEQ in 1985, it has become the primary measurement instrument of choice by sport social psychologists. Recent research has confirmed the factor structure and favourable psychometric properties of the GEQ.

Of interest relative to the measurement of team cohesion is the recent development of an instrument designed to measure team chemistry. The 53-item Team Psychology Questionnaire measures seven dimensions of team chemistry.

The seven dimensions of the TPQ are player talent and attitude, coach-technical, coach-interpersonal, task integration, social cohesion, team identity, and style of play. The TPQ follows a group cohesion model proposed by Russian psychologist Petrovsky. As proposed by Petrovsky, members of a sports team are bound together at three levels: informally as friends, formally through team roles and coordination of effort, and through a link to a greater purpose.

Determinants or Team Cohesion

Building upon Cartwright's basic model, Caron proposed a sport specific framework for studying team cohesion determinants and consequences. The basic conceptual framework is composed of four classes of determinants and two classes of consequences.

The basic notion is that there are certain factors that lead to or determine team cohesion, and certain consequences associated with having or not having team cohesion. In this section I will focus attention upon the determinants of team cohesion; in the following section, upon the consequences of team cohesion.

In an important study reported by Widmeyer & Williams, factors that determine team cohesion among female collegiate NCAA Division I golfers were investigated and reported. In this investigation, team cohesion was measured Using the multidimensional GEQ.

The strongest predictor of team cohesion, however, was personal satisfaction For intercollegiate golfers, the best way to develop team cohesion is by cultivating a personal feeling of satisfaction towards the team and team members.

Widmeyer & Williams' research is a benchmark study because it systematically studied the relationship between Carron's testimony of team cohesion determinants and team cohesion among intercollegiate golfers. Whether specific findings of Widmeyer & Williams can be generally as strong a predictor team cohesion in basketball as it is in golf?

This is an important research question, since team cohesion is generally believed to be of greater consequence in team sports than in individual sports.

Although the Widmeyer & Williams study was the most comprehensive investigation to date dealing with determinates of team cohesion, other studies have focused upon specific antecedents. For example, tea stability as an antecedent of team cohesion has been studied by Donnelly, Carron, & Chelladurain, with the general finding that team stability fosters cohesion. Similarly, research by Widmeyer, Brawley, & Carron suggests that team cohesion decreases as team or group size increase.

Consequences of Team Cohesion

Most research on consequences of team cohesion has focused upon performance. The primary question that has been asked is to what degree team cohesion leads to improved team or individual performance. Where performance is describe in terms of individual and group outcomes. Absolute and relative performance effectiveness refers to the difference between winning or losing a contest, opposed to performing better than the last time. Having a team's performance reduced to a slash in the win or loss column is an absolute measure of performance effectiveness, whereas comparing a team's performance to how well it performed in the last outing is a relative measure of performance effectiveness.

A similar dichotomy can be developed for the performance of individual sport athletes. From an absolute performance perspective, a golfer may not "win" a golf tournament, but from a relative perspective, she may have improved her score significantly.

In addition to the direct effects of team cohesion on performance, there are numerous other consequences of team cohesion that are indirectly related to performance. These include group self-esteem, future participation, lowered negative affect, team homogeneity, and self-handicapping. Each of these factors will also be briefly discussed in this section on consequences of team cohesion.

(i) Moderator of the Disruptive Effects of Self-Handicapping

Self-handicapping represents the strategies athletes use to pro-actively protect their self-esteem by creating excuses for their performance in forthcoming events by adopting or advocating impediments for success. Typical excuses might include missing practice due to injury or illness, parting and loss of sleep, school commitments or distractions, and family commitments or distractions. If success follows, the athlete or athletes can always internalize the victory, but is failure follow, they will have numerous external explanations as to why they have failed.

This behaviour causes disruption in the athlete's preparation for competition, and is therefore referred to as self-handicapping. Research indicates that team cohesion has a moderating effect on the trait of self-handicapping.

Athletes high on the trait of self-handicapping rate the severity of disruption associated with it as low when team cohesion was high. There is something about being a member of a cohesive group that makes athletes sensitive to disruptions associated with self-handicapping.

(ii) Predicting Future Participation

For young athletes especially, it is important that the sport experience lead to the expectation of continued participation. Sport participants who exhibit high levels of social cohesion also exhibit high scores in the expectation that they will participate in sport during the following season. Thus social cohesion is a predictor of the intention to continue sport involvement.

This prediction is undoubtedly related to further observation that high levels of team cohesion are related to lower state anxiety individuals low in state anxiety are more likely to continue sports participation.

(iii) Direction of Causality

As was briefly mentioned above, numerous investigations have verified that a significant and positive relationship exists between direct measure of team cohesion and performance in both individual and team sports. The issue of direction and performance in both individual and team sports.

The issues of direction of causality, however, has been a difficult issue to resolve. Does team cohesion lead to or cause successful performance, or does successful performance lead to or cause successful performance, or does team successful performance lead to or course high team cohesion? As you might guess, this is not a good "either/or" question. It is likely that high team cohesion leads to high performance, but it is also likely that successful performance leads to perceptions of team cohesion.

The critical issue is which direction is the most dominant. We would like to think that the direction from team cohesion to successful performance is most dominant almost all athletes, however, have experienced the "halo effect" of success. When teams are winning, it is a lot easier to feel at one with your team and with your team-mates.

Early research on the causality issue generally supported the position that going from successful performance to perceptions of team cohesion yielded a stronger predictive relationship than going from perceptions of team cohesion to improved performance.

The limitation of these early studies, however, was that they were all conducted using measurement instruments that tended to be unidemensional and focused to some degree upon indirect measures of cohesion. Since the development of the multidimensional Group Environment Questionnaire in 1985, few studies have addressed the causality issue, An exception to this general observation is an important study by slater and Sewell.

Slater and Sewell utilized a cross-lagged correlational design to study the causality relationship between team cohesion and performance. This particular correlational design was also used in several of the early studies on causality.

The reader is referred to Cox and to Bakeman and Helmreich for detailed explanations of this design. As the title of this technique indicate, CLCD is based on measuring team cohesion and team performance once near the beginning of the competitive season, and once near the end of the competitive season.

This configuration allows the researcher to compare correlation coefficients leading from early team cohesion to late performance to those that lead from early performance to late cohesion.

Utilizing male and female collegiate hockey team

as, Slater and Sewell demonstrated that team cohesion measured early in the season predicted late-in season hockey performance much more strongly than performance measured early in the season predicted late-in-season team cohesion.

The unmistakable conclusion from this research is that the primary direction of causality is from team cohesion to performance, and not from performance to team cohesion.

3

THE CHARACTERISTICS, SCOPE AND METHODS OF PSYCHOLOGY

Psychological Views

Perhaps a useful way of illustrating important shifts in how psychologists have defined their field is to imaging a brief conversation between three major figures in the history of the field: Wilhelm Wundt, the founder of the first psychological laboratory; William James, author of an early influential text, Principles of Psychology; and John B. Watson, founder of an approach that dominated the field until well into the twentieth century. It should be noted that this conversation is purely imaginary.

Today most psychologists believe that overt behaviour is indeed one of the central concerns of their field. They also recognize, however, that it is definitely not its only concern. In order to obtain complete and accurate answers to the complex questions with which psychology grapples, it is necessary to consider the perspectives as well. The most important of these are described below.

The Development of Psychology

In movies and television shows, scientists are sometimes represented as magnificent loners—geniuses who work in isolation and develop major break through out of their own creative spirit. While breakthroughs do sometimes occur, more frequently, scientific progress flows naturally from what went before. Modern psychology is no exception to this general rule. When psychology emerged in the late 1800s as an independent field of study, it had important roots in several other disciplines, ranging from philosophy on the one hand to biology and psychology on the other. Given these diverse origins, it is not surprising that disagreement over the nature of the new science soon developed. Early psychologists did not always agree about what science soon developed. Early psychologists did not always agree about what their field should study or how their efforts to understand various topics should proceed. Since these disputes played an important role in the history of psychology and contributed to the definition offered below, we will discuss them briefly.

The Cognitive Perspective

Consider the following incident. You have been asked to write a letter of recommendation for someone. How do you proceed? Probably, by thinking about this individual and bringing many incidents to mind. From these you try to extract some basic themes. Notice that you do not select just any themes or bring just any incidents to mind. Rather, since this is a letter of recommendation, you tend to focus on positive experience and events—ones that will help you to write a good, supportive letter. If for some reason you have great difficulty in remembering favourable incidents, your friend is in serious trouble! In any case, since you can include in the letter only information you recall, the workings of your memory are crucial to carrying out the task favourably.

This and countless other examples demonstrate that if we wish to understand people and how they behave, we must pay careful attention to *cognitive processes*—the ways in which people think, remember, decide, and so on. Indeed, many would argue that attempting to understand behaviour without attention to cognitive processes is akin to trying to solve a complex jigsaw puzzle without all the pieces.

The Physiological Perspective

What happens inside your body when you think, listen to music, grow hungry, or become angry? And what takes place when you dream, experience anxiety, become sexually aroused, or simply read a text like this one? Obviously, something must be occurring in all these cases; we are living beings, and every experience we have must be accompanied, at some level, by physiological events. Activity in our brains and other parts of the nervous system, hormones released by our glands and the bodily changes these include—such processes are intimately linked to everything we do, think, feel, or say. Psychologists have long recognized this fact and generally agree that understanding these biological roots is an essential component of the field.

Psychology: A Working Definition

So where does all this leave us with respect to working definition of modern psychology? Taking these various perspectives (behavioural, cognitive, physiological, sociocultural, psychodynamic) into account, it seems appropriate to define *psychology* as the science of behaviour and cognitive processes—the field that seeks to obtain scientific information on everything we think, feel, or do. Further, this definition implies that modern psychology attempts to accomplish this task by examining overt behaviour, cognitive processes, physiological events, social and cultural influences, hidden and largely unconscious processes, plus the complex interactions among these different factors. No wonder we do not yet have complete or final answers to many of our questions!

The various approaches exist in isolation. On the contrary, these perspectives are actually general themes adopted in all areas of the field. For example, consider love—the topic with which we began. Clearly, love involves overt behaviour; people who are in love act differently from those who are not in love. They spend long periods of time gazing at the object of their devotion, and expend tremendous effort in trying to please this person. But love also involves physiological processes—indeed, we can readily measure the physiological changes induced by the presence of a person's love and the bodily changes that occur during lovemaking.

Can love also be studied from a cognitive perspective? Absolutely. Thoughts or memories about one's lover can often generate profound bodily reactions. People often think long and hard about the question of whether they are really in love or merely infatuated with someone they have met. And their decisions in this regard strongly affect their overt behaviour. Does love involve hidden motives and the impact of unconscious forces or processes? Again, many psychologists would answer yes, and might add that such factors, too, play an important role. Finally, there can be little doubt that social and cultural factors play a role. In some societies the idea of romantic love is virtually unknown. People in such cultures report physical attraction to others, but they do not describe themselves as being in love, and they do not behave in ways persons from cultures that do possess an idea of love might expect. For example, they willingly participate in arranged marriages. In sum, these five perspectives—behavioural, cognitive, physiological, sociocultural, and psychodynamic—can often be applied to understanding a single aspect of behaviour.

The Psychodynamic Perspective

Do dreams have any meaning? Why do people often experience slips of the tongue, saying for example, "I could have killed him" when they meant to say "I could have kissed him"? What accounts for bizarre forms of behaviour often described by such terms as psychological disorders or mental illness? According to the *psychodynamic perspective*, these and many other puzzling aspects of behaviour stem from continuous, and largely unconscious, struggles among hidden forces deep within our personalities. The most famous advocate of this perspective, of course, was Sigmund Freud. However, even psychologists who disagree strongly with Freud's theories of human personality and mental disorders accept the notion that behaviour is often affected by forces, urges, and tendencies largely outside our conscious recognition. Consideration of these, too, is a part of modern psychology.

The Socio-Cultural Perspective

Consider two diplomats engaged in complex negotiations over a trade treaty. One is from the United

States, the other from Japan. The American negotiator offers a proposal. Her Japanese counterpart smiles but does not give a concrete answer. The American repeats her position, and again the Japanese negotiator merely smiles. At this the American becomes angry and breaks off the negotiations. What's going on here? One answer involves the fact that these two persons come from very different cultures. Further, these cultures off contrasting guidelines to negotiators about how to behave. American culture suggests that the best approach is to get right down to business and place an offer on the table. In addition, it indicates that there is nothing wrong with rejecting an opponent's offer out-right. Japanese culture offers different messages for negotiators. It suggests that persons involved in negotiations should spend a lot of time getting to known on another before conducting business. Further, it suggests that out-right rejections are bad form—perhaps even bad manners. Thus, it is better to offer no response than a direct no.

The contrasting actions of the two negotiators in this incident illustrate an important point: In many cases, individuals perceptions, feelings, and behaviour are strongly influenced by the social and cultural systems in which they live. In order to understand fully many aspects of behaviour, we must take sociocultural factors, too, into account.

Psychology in the Twentieth Century: A Capsule History

John B. Watson unfurled the banner of a psychology focused on over behaviour as the decade of the Roaring Twenties began. Yet much important work had already been completed before the time. For example, tests designed to measure intelligence and other human characteristics had already been devised and put to particle use. And much had been learned about the functioning of our senses and the operation of human memory.

In the decades that followed, the pace of progress within psychology accelerated sharply. During the 1930s and 1940s, C. L. Hull, B. F. Skinner, and other behaviourists uncovered much new information about learning and related topics. Important work on conformity, leadership, and other aspects of social behaviour was performed during the same period. In the 1950s, the scope of psychology expanded greatly, and major advances were made with respect to knowledge about human development, motivation, functioning of the human brain, and many other topics.

The 1960s brought the emergence of many new interests and subfields such as environmental psychology—the study of how behaviour is affected by the physical world around us, and psychology and law—efforts to determine how basic psychological processes can and do affect the legal process. In addition, of course, progress continued in older and more traditional areas of research.

During the 1970s and 1980s psychology, in a key sense, came into its own. It gained increasing recognition as an active and valuable branch of science. Moreover, its scope expanded even farther, with the emergence of such new fields and specialities as adult development (development and change during our adult years), gender differences (efforts to identify and understand differences between the sexes), and health psychology (the impact of various psychological factors on physical health). Many other fields and lines of research too numerous to mention here also appeared during these exciting decades. The result is that at present, psychology has expanded its scope to the point at which it now studies virtually every question about human behaviour you can possibly imagine—and then some.

By design, this capsule overview of the history of our field is extremely brief. However, the information presented above should provide you with at least a rough idea of how our field has grown and developed during the present century.

Psychology: Who and What

Now that we've defined the field of psychology and addressed some questions about its basic nature, we'll turn to two related topics: who psychologists are and what they actually do.

Who: The Background and Training of Psychologists

The term psychiatrist and psychologist are quite similar, so it is not surprising that many persons think they mean the same thing. Actually, they refer to two different groups of professionals. Psychiatrists are physicians who, after completing medical studies, specialize in the treatment of mental disorders. In contrast, psychologists receive their training in graduate programs of psychology, where they earn both a master's degree and, in most cases, a Ph.D. The latter degree usually requires a minimum of four to five years of study. In addition, psychologists who choose to specialize in certain areas of their field, such as the treatment of psychological disorders, must also complete one or more years of practical experience in a hospital, clinic, school or business. Throughout their graduate education, psychologists focus mainly on the principles and findings of their field. However, most also complete extensive training in statistics, in research methods, and in related fields such as physiology, sociology, or management sciences.

It should be clear, then, that psychologists and psychiatrists receive different kinds of training. So why are the two fields often confused? In part because many psychologists specialize in the diagnosis, study, and treatment of psychological (mental) disorders. As a result, they focus on many of the same problems and perform many of the same activities as psychiatrists. In fact, members of the two fields often work closely together in the same mental health facilities. Since only some psychologists focus on mental disorders, the two fields overlap only partially and remain largely independent.

Now that we've clarified the difference between psychologists and psychiatrists, here are a few facts about psychologists themselves:

1. The most recent available data indicate that there are now more than 110,000 psychologists in the United States alone.

2. Of these, 61.8 percent are male and 37.8 percent are female. This represents a marked increase in the proportion of females; as recently as the 1950s only about 10 percent of all psychologists were women. Incidentally, at present, a majority of persons pursuing graduate education in psychology are female.

3. As a group, psychologists are relatively young. Almost half received their Ph.D. degree between 1970 and 1979. Thus, most psychologists are in their thirties or forties.

4. More than 3,000 doctoral degrees are awarded in psychology each year.

What: Specialities within Psychology

Before going further, stop and answer the following questions: What exactly do psychologists do? If you are like most people, you probably said "conduct therapy", "help people with problems", or perhaps "conduct research". In a sense, the first two replies are not inaccurate, for nearly half of all psychologists are clinical or counselling psychologists, and they do indeed within one mental problems and disorders. But there are many other specialities within psychology, so in reality, psychologists do many different things and investigate a wide range of topics. Here is a brief description of several of psychology's major subfields:

Clinical Psychology: Studies the diagnosis, causes, and treatment of mental disorders. For example, clinical psychologists have recently devised several effective forms of treatment for reducing serious depression.

Counselling Psychology: Assist individuals in dealing with many personal problems that do not involve mental disorders. For example, counselling psychologists assist individuals in career planning and in developing more effective interpersonal skills.

Developmental Psychology: Studies how people change physically, cognitively, and socially over the entire lifespan. For example, developmental psychologists have recently found that tendencies toward shyness may occur very early in life and are, at least in part, in inherited characteristic.

Educational Psychology: Studies all aspects of the educational process from techniques of instruction to learning disabilities. For example, educational psychologists are working to develop classroom procedures designed to help minority children in the United States overcome some of the environmental disadvantages they face.

Cognitive Psychology: Investigates all aspects of cognition—memory, thinking, reasoning, language, decision making, and so on. For example, cognitive psychologists have recently established that animals—even rats—are capable of complex forms of counting, such as keeping running counts of two types of events simultaneously.

Industrial/Organizational Psychology: Studies all aspects of behaviour in work settings—selection of employees of performance, work motivation, leadership. For example, industrial/organizational psychologists have found that individuals work harder when they have concrete goals that they accept than when they do not have specific goals.

Psychobiology (physiological psychology): Investigates the biological bases of behaviour—the role of complex biochemical events in our nervous systems and bodies in everything we do, sense, feel, or think. For exampie, psychobiologists are now attempting to identify the neural and physiological mechanisms that play a role in various types of addiction.

Social Psychology: Studies all aspects of social behaviour and social thought—how we think about the interact with others. For example, social psychologists have recently found that female leaders receive more negative and fewer positive nonverbal reactions from followers than do male leaders.

Experimental Psychology: Studies all aspects of basic psychological processes such as perception, learning, and motivation. For example, research by experimental psychologists has recently added much to our understanding of attention—the process of directing portions of our information-processing capacity to specific stimuli or events. This understanding, in turn, is now being applied to the design of more effective warnings for various hazards.

The exact percentages are not crucial to remember; rather, keep in mind two facts: (1) psychologists do a number of very different things, and (2) despite the existence of many subfields or specialities, psychology remains a unified field with shared values, goals, and overall methods.

Psychology: Some Basic Questions

Definitions, it might reasonably be stated, are made to be challenged. Given that fact, we'll now consider some of the questions you may have about the suggestion that psychology be viewed as the science of behaviour and cognitive processes.

Is Psychology Really Scientific? To start at the beginning, let's address a question sometimes heard from students: Is psychology really scientific? Even today, many persons seem to feel that it is not a science in the same sense as chemistry, physics, or biology. The reason for such belief is that people misunderstand the meaning of the word science.

Many persons assume that this term refers to specific fields of study, such as chemistry and physics, and that only these are truly scientific in nature. In fact, science actually refers mainly to a general approach to acquiring knowledge—one involving the use of certain methods plus adherence to several key values or standards. The methods consist primarily of systematic observation and direct experimentation. The standards involve commitment to such goals as objectivity (evaluating information on the basis of its merits rather than according to one's personal preferences), accuracy (gathering information as carefully and precisely as possible), and skepticism (accepting findings as true only after they have been verified over and over and all inconsistencies have been resolved).

These methods and values are largely independent on the content of any specific field. In fact, they can be used to study a wide range of topics. In determining whether a field is or is not scientific, then, the crucial question in this: Does it make use of scientific methods and accept scientific values? To the extent that it does, it may be viewed as scientific in nature. To the extent that it does not, it should be seen as basically non-scientific in its approach. In short, it is the methods and values, not the topics being studied, that are essentials.

Given the criteria just described, is psychology really scientific? The answer of this question is yes. In their efforts to understand behaviour and cognitive processes, psychologists rely heavily on scientific methods and adhere closely to the standards mentioned above. This does no imply, however, they are always perfect in such efforts. As you'll see below, in the discussion of the role of theories in psychology, psychologists—like other scientists—sometimes find their objectivity threatened by their own personal views or attitudes. However, like other scientists, they make strenuous efforts to avoid such effects and to maintain their objectivity, and this helps ensure the essentially scientific nature of the field.

To conclude: it is reasonable to define psychology as the science of behaviour and cognitive processes, because in their research psychologists are firmly committed to the scientific method and to the values of science. The topics they seek to study certainly differ from those in older and more traditional fields of science, but the approach they follow is basically the same. And that, is really the central issue.

Is Psychology Merely Common Sense? In a way, everyone is an amateur psychologist. We all think about our own behaviour, and that of others, on a regular basis. Because we do, we often feel that we already know the answers to many of the questions psychology studies and that we can predict the results of psychological research on this basis. Are such impressions correct? In other words, is psychology merely a rehashing of what our common sense and informal experience already tell us? Before concluding that this is so, consider the these two points.

First, if you stop and think about it for a moment, you'll quickly realize that common sense isn't quite as useful a guide to human behaviour as it seems. Often it yields contradictory answers about important issues. For example, according to common sense, "Absence makes the heart grow fonder". But common sense also notes, "Out of sight, out of mind". Can both be true? Perhaps, but common sense does not explain why or how. As another example, consider these two statements: "Birds of a feather flock together" (similar persons are attracted to one another) and "Opposites attract" (dissimilar persons are mutually attracted). Again, we are left with contradictory suggestions. This is where a scientific field like psychology comes into on its own. Through systematic research, psychology can help determine whether any of these commonsense suggestions are true. Going farther, it can also find out why such effects occur—why, for example, the more similar two persons are, the more they tend to like each other. This is actually the case: there is very little support for the view that opposites attraction.

Second, before concluding that many of the findings of psychology are merely common sense, you should consider a phenomenon known as the hindsight or "I-knew-it-all-along" effect. This refers to our tendency to assume, on learning of some event, that we knew it would happen all along and therefore it is not all that surprising. Why do we perceive that new information is not really new and that, in fact, we knew it (or at least would have predicted it) all along, this confirms the vast scope of our personal wisdom. Because of the hindsight effect, people often perceive the results of psychological research as something they would have predicted if asked in advance. In fact, however, several studies indicate that when people are asked to predict the results of actual research, they are often wrong. So, before you decide that the knowledge gathered by psychologists is usually something you really knew all along (or something your grandmother could have told you if you asked), remember that hindsight is often a lot better than foresight.

Adding to What We Know: The Process of Psychological Research

Now let's turn our attention to what is perhaps the most importan⁺ question in sport psychology. How do psychologists perform the task of adding to our knowledge about behaviour and cognitive processes? How do they move beyond common sense to answer puzzling questions about human behaviour? Psychologists use several techniques for conducting systematic research—the scientific method of seeking answers. In this section we will discuss the most important of these procedures and also consider some of the complex ethical issues they sometimes raise.

Naturalistic Observation: Scientific as Explorers

Almost everyone finds the giant panda of China fascinating. Here, at least in outward appearance, is a teddy bear come to life. For many years zoos throughout the world sought eagerly to add these appealing animals to their collections. Unfortunately, these efforts usually produced disappointing results. The pandas, when finally obtained, seemed to pine away in captivity.

Even worse, in many zoos they adamantly refused to mate, despite the best efforts of the experts and panda's keepers. Given the small and declining number of pandas present in the wild, these events seemed to spell disaster for a species most people would very much like to preserve. Could anything be done to change this situation? There seemed only one way to find out: observe the pandas in their natural habitat to learn more about their behaviour seemed to be a situation in which research should be conducted through *naturalistic observation*—systematic study of behaviour in natural settings. Fortunately, efforts along these lines have added greatly to our knowledge of giant pandas. Scientists have spent months observing the pandas in the mountainous regions of western China where they live, and from such study have extracted information that may prove useful in assuring the species survival.

As this example suggests, naturalistic observation is often used to study animal behaviour. However, it is sometimes applied to human beings as well. For example, in a project we'll consider later in more detail, Murdoch and Phil observed the behaviour of male patrons drinking beer or liquor in randomly selected bars. Information on the number of drinks they consumed and resulting changes in their reactions to others offered intriguing insights into the effect of alcohol on human behaviour in at least one setting.

Sometimes, the data obtained though naturalistic observation are relatively informal, and this can reduce their scientific value. However, the fact that subjects are studied in natural settings and so are likely to act in the ways they normally do is an important advantage that makes this method useful in some contexts.

Case Studies: Generalizing from the Unique

Each human being is unique—each possesses a distinctive combination of traits, abilities and characteristics. Given this fact, it is possible to learn anything about human behaviour generally from detailed study of one or perhaps a few persons? Several famous figures in the history of psychology have contended that it is. Thus, they have adopted the *case method*, in which detailed information is gathered on specific individuals. For example, Freud based his entire theory of personality on the case method. Is the case method really useful? In the hands of talented researchers such as Freud, it does seem capable of yielding valuable insights about behaviour. Moreover, when the behaviour involved is very unusual, the case method can be quite revealing. Detailed study of several unique cases has added greatly to our understanding of the biological bases of memory. These cases involved individuals who have experienced specific kinds of damage to the brain and, as a result, show certain kinds of memory deficits. By studying the pattern of such memory losses, scientists have been able to piece together a more complete picture of how memories are stored in the brain. So, emotional attachments to people with whom they work closely for months or even years can reduce their objectivity—the case method does appear to have its uses. When used with considerable caution, it can prove helpful in the investigation of at least some aspects of behaviour.

Surveys: The Science of Self-Report

The *survey approach* is another research method used by psychologists. This involves asking large numbers of individuals to complete questionnaires designed to yield information on specific aspects of their behaviour or attitudes. Such surveys, on polls, are often conducted to measure attitudes toward a wide range of issues, such as the reunification of Germany, abortion, or steps designed to reduce global warming produced by the greenhouse effect. In addition, surveys are employed to assess voting preferences prior to elections, consumer reactions to various products, and current patterns of behaviour for large numbers of persons (for example, health practices or compliance with safety regulations such as mandatory seat belt use). Surveys are often repeated over long periods of time in order to track shift in public opinions or actual behaviour. For example, some surveys of job satisfaction—individuals attitudes toward their jobs—have continued for several decades. And changing patterns of sexual behaviour and sexual attitudes have been tracked by the Kinsey Institute for more than forty-five years.

The survey method offers very real advantages: large amounts of information can be gathered with relative ease, and shifts over time can be readily noted. However, the disadvantages are also quite apparent. People may fail to respond accurately or truthfully, providing answers that place them in a favourable light rather than ones that reflect their true views. In addition, the results of surveys are useful only if the persons questioned are truly representative of larger groups to whom the findings are to be generalized. For example, imagine that a survey conducted with 10,000 men and women indicated that more than 90 percent were strongly in favour of a large increase in gasoline taxes to provide greater support for local school systems. Should you prepare to dig deeper into your pocket each time you pull up to the gas pump? Perhaps. But not if you learn that all 10,000 persons who responded to the survey are members of a national teacher's association! The fact that they are raises serious questions about the extent to which they are representative of the larger group to whom we wish to generalize—the entire adult population. The moral is clear: Unless the people who respond to a survey are similar to the larger group to whom we wish to extend the results, such generalization are on very shaky grounds.

The Correlational Method: The Search for Relationships

Late each summer, many television news programs present the following scene: A newscaster is shown standing in a field with a farmer. Both are examining a caterpillar. The newscaster asks the former whether, from the length of "hairs" on the caterpillar's body, he can predict the severity of the oncoming winter. The farmer answers that he can, and proceeds to do so. Supposedly, the longer the hairs on the caterpillar, the worse the oncoming winter will be.

What do these forecasts have to do with psychological research? In one sense, quite a bit. Prediction—the ability to forecast future events from present ones—is an important goal of many branches of science, including psychology. Consider, if you will, how useful it would be if we could predict from currently available information about individuals (such as test scores) future events like success in school or in various occupations, effectiveness as a parent, or the development of serious mental problems. How can this goal be attained? One answer involves efforts to determine whether various aspects of the world (termed variables because they can take different values) are related to one another. That is, we try to determine whether the changes in one variable are associated with changes in another so that, for example, as one rises, the other does too. The stronger such relationships (correlations), the more successfully one variable can be predicted from the other. Appendix A provides more information about correlations and how they are computed.

The TV scene described above is based in part on the principle of correlation. The farmer and newscaster assume that the shaggier the caterpillar's coat, the colder the oncoming winter. Hence, the second variable (winter weather) can—supposedly—be predicted from the first (the caterpillar's coat). Needless to say, the reality of this particular correlation is doubtful at best. Some events are indeed closely related to others and so can be used as effective predictors. For example, to return to weather, meteorologists (scientists who specialize in the prediction of weather) have found that the greater the number of disturbances on the face of the sun, the more unsettled the world's weather will be in the coming month. Similarly, it has been observed that the greater the number of hours of violent television shows children watch, the greater their likelihood of behaving aggressively as teenagers.

Of course, in many cases, accurate predictions can be obtained only when several factors (and the correlations between them) are simultaneously considered. For example, if you wished to predict the extent to which specific persons will be able to the stress of working on a swing-shift schedule (moving repeatedly from day shift to night shift and back again), you would probably have to consider several different variables and the relationships between them, including their current sleep habits, various aspects of their physical and psychological health, and several features of their personalities. Only if you took all of these variables into account could you make accurate predictions about their future reactions to swing-shift work schedules.

How do psychologists use correlations in their research? How do they search for relationships among variables in order to predict one or more from the others? Perhaps the best way of illustrating the nature and value of this approach—known as the *correlational method of research*—is through an actual example.

The Correlational Method: An example. Imagine that a researcher wished to test the following *hypothesis* (an as yet untested—but testable—prediction about some aspect of behaviour): the more alcohol individuals consume, and hence the higher the concentration of alcohol in their blood, the more aggressive they become. At first glance, you might assume that this is a very straightforward question, with an obvious answer: of course people become more aggressive as they drink alcohol. In fact, however, informal observational yields mixed evidence on this score. While alcohol does indeed seem to be implicated in many violent episodes, it also appears that in some situations alcohol acts as a social lubricant, putting people in a good mood and so actually reducing the likelihood that they will become irritable and lose their tempers. Here, then, is a case where systematic research is needed to clarify a confusing and contradictory picture offered by common sense.

How might a researcher who wished to use the correlational method proceed? One possibility is that he or she could observe the amount of alcohol individuals consume (one variable) and the amount of aggression they show toward others (a second variable). Then, through appropriate statistical procedures, the strength of any relationship between these two variables can be determined. A correlations can range from -1.00 to +1.00. The greater a correlation's departure from 0.00, the stronger the relationship between the variables being considered. Remember: the stronger the correlation, the more accurately each variable can be predicted from the other.

Procedures very similar to these were actually employed by Murdoch and Pihl in an ingenious field study. These researchers visited many different bars in a large Canadian city and gathered data on several factors that they believed might be related to aggression in such settings. Specifically, they approached male patrons of the bars, identified themselves as researchers, and then asked these persons questions about how many drinks they had consumed, the time of their first drink, what they were drinking, and how intoxicated they felt at the moment. In addition, the researchers also gathered information on participant's weight, height and age.

While aggression is hardly rare in bars, the researchers wanted to observe its occurrence under standardized conditions. Thus, they arranged for an assistant to approach each participant and ask a series of questions: "Do you have the time?" "Would you have change for a dollar?" "Would you have a piece of paper I could write a message on?" The idea was to provide all participants with uniform opportunities to lash out verbally against a stranger. Replies to the assistant's questions were rated on a special aggressiveness scale.

For example, a polite refusal to answer further questions constituted a low level of aggression. A refusal in a sarcastic tone represented intermediate aggression. And verbal comments reflecting a loss of patience or a threat, like "You're pushing your luck, buddy", represented high levels of aggression.

Results were intriguing—and surprising. First, it appeared that persons drinking liquor were slightly more aggressive, overall, than those drinking beer, even when the amount of alcohol in their bloodstreams was equal. Second, aggression increased as the assistants asked repeated questions and however, was the following result: Neither the number of drinks consumed by participants nor the estimated alcohol content of their blood was significantly related to level of aggression. In other words, correlations between these variables were not significantly different from 0.00. So, contrary to what common sense might predict, it was not the case in this study that more participants drank, the higher their level of aggression.

The provocation results of this investigation help to highlight and advantages of the correlational method of research. First, this approach is often quite efficient, and it can be used to gather data about almost any variables an investigator wishes to study. Second, it can be employed in naturalistic settings such as the bars visited by Murdoch and Pihl. Third, it can be used to study many variables at once—a necessary if we wish to be able to accurately predict complex forms of behaviour. Finally, because of practical or ethical considerations, it is often the only method of research that is feasible. In other words, the only approach possible is to observe variables of interest to see if they are related in any way. Unfortunately, though, the correlational method suffers from one drawback that lessens its appeal, at least to a degree: the findings it yields are not conclusive with respect to cause-and-effect relationships. That is, the fact that two variables are correlated (that changes in one are accompanied by changes in another) does not guarantee that there is a causal link between them—that changes in the first cause changes in the second.

Correlation Versus Causation

To see why correlation does not necessarily imply causation, let's return to the question of whether alcohol consumption is related to aggression. Imagine that in their study, Murdoch and Pihl had found a moderate correlation between the amount of alcohol consumed by customers and their tendencies to act aggressively. Does this mean that alcohol consumption causes people to become aggressive? Not necessarily. It is possible, for example, that people who drink a lot in bars are different, in some respect, from people who drink smaller amounts. For example, they may be individuals who are relatively uninhibited or who seek excitement. Thus, it might be these factors—not alcohol itself—that account for the individual's higher level of aggression. In other words, the correlation between alcohol consumption and aggression does not imply that one causes the other. In fact, both high levels of drinking and high levels of aggression may stem from one or more other variables that underlie both.

Perhaps an additional example will help to clarify the reasons why even a high degree of correlation between two variables does not necessarily mean that they are causally related. First, imagine that we measured both the weight and the salaries of a large group of executives. What might we find? One possibility is a positive correlation between these variables—the higher the weight of these persons, the higher their salaries. Does this mean that weight gain produces higher income? Hardly. Both variables are related to a third—age. As people grow older, many tend to gain weight; and as people age, they advance in their careers and obtain higher salaries. Alternatively, weight and salary may rise together because when people work the long hours necessary for success, they have little spare time for exercise! So sometimes a positive correlation between two variables merely means that both are related to another variables—not that changes in one cause changes in the other.

In view of these example, the main point should be clear. The existence of even a strong correlation between two factors should not be interpreted as a definite indication that they are causally linked. Such a conclusion is justified only if it is supported by additional evidence ruling out the possibility that the correlations derive from additional factors not taken into account.

Experimentation: Knowledge Through Intervention

While psychologists use all of the research methods described so far, many prefer the *experimentation* approach. Experimentation involves efforts to determine if variables are related by systematically varying one (or more) and observing the effects of such variations on the other (or others). There are several reasons why psychologists prefer this basic approach, but perhaps the most important is this: in contrast to the other methods we have considered, experimentation yields relatively clear-cut evidence on causality. If systematic variations in one factor produce changes in another (and if additional conditions we'll soon consider are also met), we can conclude with reasonable certainty that there is a causal link between them. Establishing such causality is extremely valuable from the perspective of one major goal of science: explanation. Briefly, scientists do not wish merely to describe the world around them and the relationships between different variables or factors in it. They also wish to be able to explain why such relationships exist—why, for example, people in a good mood are sometimes less likely to take risks than people in a neutral mood; why person with certain personalities are more likely than others to suffer heart attacks; why some persons but not others suffer from deep depression. Experimentation, because it often yields information useful in answering such questions, is frequently the method of choice in psychology. It should be noted: there are no hard-and-fast rules in this regard. Rather, most psychologists select the research technique that seems most suited to the topic they wish to study and the resources available to them.

Unfortunately, many persons seem to view experimentation as both mysterious and complex. In fact, it is neither. The basic logic behind it is surprisingly simple. To help you understanding its use in psychological research, basic nature of experimentation should be understand—how it actually proceeds. Then we'll comment briefly on two conditions essential for its success. Finally, we'll describe some potential pitfalls that can, if overlooked, cast doubt on the validity of experimental findings.

Experimentation: Its Basic Nature

Reduced to its bare essentials, the experimentation method in psychology involves two basic steps: (1) the presence or strength of some variable believed to affect behaviour is systematically altered, and (2) the effects (if any) of such alterations are measured. The logic behind these steps is as follows: if the factor varied does indeed influence behaviour or cognitive processes, then individuals exposed to different levels or amounts of that factor should differ in terms of their behaviour. Thus, exposure to a small amount of the variable should result in one level of behaviour, exposure to a larger amount should result in a different level, and so on.

The factor systematically varied by the researcher is termed the *independent variable*, while the aspect of behaviour or cognitive processes studied is termed the *dependent variable*. In a simple experiment, then, different variables (such as low, moderate, and high). The participants behaviour is then carefully measured to determine whether it does in fact vary with different levels of the independent variable. If it does—and if two other conditions described below are met—the researcher can tentatively conclude that the independent variable does indeed cause changes in the aspect of behaviour being studied.

Perhaps a concrete example will help clarify this process. Let's return to the possible effects of alcohol on aggression. How could this issue be studied by means of the experimental method? Here is one approach. The independent variable in this research—the variable believed to affect behaviour—is the amount of alcohol consumed, and thus the level of alcohol in the bloodstream. The dependent variable is overt aggression. A researcher studying this issue could arrange for different groups of participants to drink different amounts of alcohol (low, moderate, high). Then the researcher could measure the tendency of individuals in these different groups to lose their tempers and agrees against others. In fact, such procedures have actually been used in several recent experiments. In one such study, Taylor and Sears gave male participants either a large dose of alcohol (drinks containing 1.5 ounces of alcohol per forty pounds of body weight) or no alcohol (drinks that smelled of alcohol but actually had none in them). This latter group served as a *control group*—persons who are not exposed to the independent variable and so provide a baseline for comparison with subjects exposed to this factor. It is to be noted that in research such as that conducted by Taylor and Sears, only individuals who are willing to consume alcoholic beverages participate. Persons who object to consuming alcohol would not, of course, take part.

After consuming alcoholic or nonalcoholic drinks, participants competed with another person in a reaction-time task. In this task, the slower player on each trial would supposedly receives a shock set for him by his opponent. Thus, participants willingness to agrees against the opponent could be measured in terms of the strength of the shock levels they chose. In reality, participants could not actually harm another person—they merely believed that they could. Would individuals who consumed high doses of alcohol demonstrate higher aggression toward their opponents? In fact, they did not. When permitted to choose the shock settings without interference, those in the high-dose group did not choose significantly higher shocks than those in the no-alcohol control group. However, another part to the study introduced a second independent variable. During the early phases of the competitive game, two other persons (assistants of the researcher) were present in the room and urged participants to deliver strong shocks to their opponents. In the presence of such pressure, persons who consumed high doses of alcohol were indeed more aggressive. This pattern of findings indicates that there was an interaction between two variables: the effects of one (dose of alcohol) depended on the presence of the other (social pressure to behave aggressively). Because variables that influence behaviour often interact in this manner, psychologists frequently include two or more independent variable in their experiments. The results obtained from such studies are often more informative than those obtained from studies in which only one independent variable is systematically altered.

Experimenter Effects and Demand Characteristics: How to Get Positive Results without Really Trying

Before concluding this discussion of experimentation, we should consider two additional potential pitfalls lying in wait for careless researchers. The first of these is known as the risk of *experimenter effects*—the face that sometimes, researchers can influence the behaviour of participants without intending to do so. Experimenter effects can occur in several ways. First, researchers usually have expectations about how participants may behave. These expectations may influence their behaviour toward participants and so alter the results obtained. For example, imagine that a researcher studying the impact of alcohol on aggression believes that high doses of this drug will in fact make people more aggressive. He or she may then nod, smile, or otherwise show approval when individuals in the high-dose condition act aggressively, but may fail to show such behaviour when those in the no-alcohol control group do so. The result: Participants may notice these reactions and be affected by them.

The second potential snare involves the fact that experimenters can sometimes unintentionally communicate the hypothesis behind the study to participants. Once they do, participant's behaviour may be affected even if the researchers have no direct contact with them and cannot provide subtle cues of approval such as those described above. The effects of such communication are known as *demand characteristics*, since they place subtle demands on participants to "help" the researcher by confirming his or her hypothesis. Imagine that in a study on alcohol and aggression, the researcher reveals the hypothesis while giving initial instructions to participants. These persons may then feel obligated to confirm the hypothesis by acting in ways consistent with it. For example, if the researcher indicates that the hypothesis behind the study is that people given a high dose of alcohol will not differ from those given no alcohol, participants in the high-dose group may bend over backwards to avoid behaving aggressively, and so help confirm the experimenter's hypothesis. Of course, participants are not always so kind: sometimes if they discern the hypothesis being examined they may do the opposite—perhaps to show that they are not so easily predicted.

Research indicates that experimenter effects and demand characteristics can both exert powerful influence on participant's behaviour. Thus, it is crucial that these effects be minimized in all psychological research. One way of doing this is through a *double-blind procedure*. Here, persons who have contact with participants (often research assistants) are unfamiliar with the hypothesis under investigation and do not know the condition to which participants have been assigned. Under these conditions, they cannot readily communicate the hypothesis being studied and cannot have clear expectations about how a given participant "should" behave. Another technique for avoiding experimenter effects and demand characteristics is to minimize direct contact between participants and researchers. This can be accomplished through the use of computerized procedures, in which participants receive instructions and perform experimental tasks by means of a computer terminal. Through these and related procedures, the impact of several potential pitfalls can be reduced and the validity of accuracy of experimental findings enhanced.

Experimentation: Two Requirements for Its Success

Before we can conclude from experimentation that an independent variable has caused some change in behaviour, two conditions must be met. The first condition involves *random assignment of participants to experimental conditions.* According to this principle, all participants in an experiment must have an equal chance of being exposed to each level of the independent variable. The reason for this rule is simple: if participants are not randomly assigned to each group, it may later be impossible to determine whether differences in their behaviour stemmed from differences they brought with them to the study or from the effects of the independent variable itself. For example, consider the study on alcohol and aggression. Imagine that two groups of men participated: male college students from a sociology class and male college students from a karate class. Because of scheduling problems, all those given high doses of alcohol belong to the karate class and all those given no alcohol from the sociology class. Now we find that those in the high-dose group are indeed more aggressive than those in the control group. Why? In fact, we cannot tell. It may be that high doses of alcohol make people more aggressive. Alternatively, it may be that persons who take karate classes are more aggressive to start with. It is to avoid such problem that the random assignment to conditions rule is followed in psychological research.

The second condition essential for conducting a successful experiment is this: Insofar as possible, all factors that might also affect participants behaviour, aside from the independent variable, must be held constant. To see why this is so, imagine that the alcohol/ aggression study is conducted over a period of several days. The no-alcohol control group participants first, and as much would have it, the weather is beautiful during this period: cool and sunny. When participants in the high-dose condition are tested, however, the weather is hot, muggy, and stifling. Again, assume we find a difference between the two conditions: those who receive the high dose of alcohol are more aggressive. Why? Again, we cannot really tell. It might be the impact of the alcohol, or it might be the influence of the weather:

research findings indicate that people are indeed more likely to be irritable and aggressive when the weather is uncomfortably hot than when it is pleasantly cool. In such cases, the independent variable is said to be confounded with another variable; in other words, the effects of each on the aspects of behaviour being studied cannot be separated or disentangled. Only when such confounding is prevented can the results of an experiment be interpreted with confidence as indicating a causal link between the independent and dependent variables.

The Role of Theory in Psychological Research: Major Benefits, Potential Dangers

Now that we have considered basic aspects of the research process in psychology, we can turn to a question students often ask: "Just how do psychologists come up with the ideas with the ideas for all those studies anyway?" As you can probably guess, there is no single answer, for several factors play a role. Some research projects are suggested by informal observation of the world around us. Psychologists take note of some puzzling pattern of behaviour or cognition and plan investigations to increase their understanding of that pattern. At other times the idea for a research project is suggested by the findings of earlier studies. Successful experiments do not simply answer questions; they often raise additional ones as well. Thus, the problem facing psychologists is usually not that of coming up with interesting ideas for research. Rather, the difficulty is in choosing among the many intriguing possibilities. Perhaps the most important single basis for research is psychology, however, is formal *theories*.

Theories represent efforts by scientists in any field to answer the question "Why?" In more formal terms, theories consist of two major parts: (1) basic concepts, and (2) statements concerning relationships between these concepts. Scientists use these parts to generate testable propositions—ones that can be examined in actual research and found to be either true or false. The development of accurate theories is a major goal of all sciences, and psychology is no exception. Psychologists, like other scientists, wish not merely to observe or describe the phenomena or aspects of behaviour they study; they wish to explain them as well. Thus, a great deal of research in psychology is concerned with efforts to construct, refine, and test specific theories. To illustrate the central role of theory in ongoing research, let's turn to another concrete example.

Consider the following: when individuals make a decision, they tend to stick to it even in the face of growing signs that it was wrong. For example, they remain in romantic relationships even after these continue to yield more pain than joy. Or they continue to "throw good money after bad" by repairing an automobile that requires one repair after another. This pattern is sometimes known as escalation of commitment.

Certainly, knowing about this tendency is useful in itself. It allows us to predict what may happen in situations where people invest resources or effort in failing courses of action: all things being equal, they will continue to stick to their initial bad decisions. Further, it also suggests the possibility of intervening in some manner to prevent such outcomes. For example, informing people about this tendency to throw good money after bad may prove helpful in lessening its occurrence. These two outcomes—prediction and intervention—are major goals of science. Knowing about escalation of commitment, however, does not explain why it occurs. This is where theory enters the picture, for, as noted above, providing such explanation is the foremost goal of theories.

Interpreting Research Results: Statistics as a Tool for Evaluating Data

Once an experiment has been completed, researchers must turn to the next crucial task: interpreting the results. The best means of illustrating this important process is, again, through a concrete example. Imagine that a psychologist conducts an experiment designed to shed light on the following question: Do people remember information consistent with their current moods more readily than information inconsistent with these moods? Specifically, do people in a good mood remember positive information more readily than negative information while people in a bad mood remember negative information more readily than positive information? To test this hypothesis, the psychologist asks participants to memorize a list of words, some of which have a positive meaning (good, happy, success) and some of which have a negative meaning (bad, unhappy, failure). Then the researcher induces either a positive or a negative mood among participants by exposing them to praise or criticism of their words. Finally, participants in both groups are asked to remember the list of words. Results seem to support the hypothesis: Participants in a good mood remember more positive than negative words, while those in a bad mood remember negative than positive words. How much confidence should the psychologists place in these findings? In other words, are the differences between the groups real—ones that would be observed again if the study were repeated with other groups? This is a crucial question, for unless we can be confident that such replication would occur, an experiment's results really tell us little about behaviour.

One way of dealing with this question of course, would be to actually repeat the study over and over again, but as you can well imagine, this would be costly in terms of time and effort. Another approach involves the use of *inferential statistics.* This is a special form of mathematics designed, in part, to evaluate the likelihood that a given pattern of findings, such as differences between experimental groups, is due to chance alone. Thus, to determine whether the findings of a study are indeed real—that is, unlikely to have occurred by chance alone—psychologists perform appropriate statistical analyses on the data they collect. If these analyses suggest that the likelihood of obtaining the observed findings by chance is low—usually, fewer than five times in a hundred—the results are described as being significant. And only then are they interpreted as being of value in the task of understanding some aspect of behaviour.

It should be noted that the likelihood that a given patterns of finding is a chance events is never zero. This probability can be a very low—for example, one chance in ten thousand or even one chance in a million—but there is always some possibility, however small, that a pattern occurred by chance. For this reason, actual replication of results by different researchers in different laboratories is usually necessary before the findings of any research project can be accepted with confidence, even if statistical analysis indicates that they are unlikely to be a chance event. Still, inferential statistics is a valuable tool of tremendous help to psychologists in evaluating research findings.

In sum, because of their commitment to scientific skepticism, psychologists are reluctant to accept the results of any research project—even their own—as valid. Only when the data have been subjected to appropriate mathematical (statistical) analysis and significant effects have been obtained do psychologists view experimental findings as useful in understanding various aspects of behaviour.

4

MOTIVATION AND SPORTS BEHAVIOUR

Motivation is anything that impels an organism to action. Action is reflected in the selection, intensity, and persistence of behaviour. Drive theory was the earliest and most traditional approach to describing the factors underlying motivated behaviour. However, when the premises of drive theory were found to be somewhat lacking, scholars turned to more cognitive, or organismic, explanations. They began to explore questions such as "How do an athlete's thoughts affect performance?" "How do emotions relate to motivation?" "In addition, two important themes evolved in the sport motivation literature. The performer's perception of self-competency and personal control over outcomes appear to be directly related to motivation.

ATTRIBUTION THEORY

The focus of attribution theory is the relationship between the explanation a person gives for performance outcomes and effects on subsequent actions. Those explanations rest on a person's perception of causality. Thus attribution theory is a cognitive approach to motivation because it assumes that people attempt to understand and explain events to provide themselves with a rational basis for their own behaviour.

Instead of using complicated formulas and constructs to predict behaviour, attribution theory uses the vocabulary of the layman; hence, it is referred to as native psychology. The attributions a performer uses to explain either success or failure in sport settings can ·influence both expectancy of future performance outcomes the emotions the performer experiences regarding the outcome in question. For example, suppose that a high jumper has a particular excellent jump that exceeds his personal best.

When asked to explain why he thinks he did so well, he replies that he really tried hard. He is then attributing the successful outcome to effort. This attribution should have two associated effects. First, the high jumper should perceive that he can repeat the successful performance in the future and will most likely be motivated to do so.

Second, by attributing the successful outcome to an internal factor, effort, the accompanying effect should be feelings of pride. But suppose the jumper attributed the same outcome to lack. Will the effects on motivation and affect be identical? No. Luck is an external factor not under the performer's control. It may or may not be present on a future occasion. In addition, there would be little reason to experience high degrees of positive affect concerning a performance accomplishment that occurred because an element such as luck was on your side. Clearly, causal attributions used by a performer can influence both expectancy of future performance outcomes and affect.

Causal Elements

Causal elements refer to the reasons people give for experiencing either success or failure. Although there are many causal explanations performers could evoke, such as officiating bias or fatigue, the attributions usually studied are those made to ability, effort, task difficulty, and luck. The two most important causal elements are the internal factors of ability and effort.

Expectancy and Affect

The stability dimensions is most closely associated with future expectancy of success or failure. By attributing a positive outcome to a stable factor such as ability, there is every reason to believe that the same outcome will recur. Conversely, attributing a failure outcome to lack of ability should foster the perception that future failure is inevitable. If that stable attribution was used in a consistent fashion, then it could lead to impaired motivation based on feelings of incompetency and perhaps learned helplessness.

Learned helplessness is a perception of "uncontrollability", an independence between a person's responses and their consequences. According to Weiner's original model, locus of causality is predicted to be most closely associated with affective responses. Specific feelings apparently are generated as a result of attributing outcomes either internally or externally.

Notice that internal attributions for success tend to enhance feelings of self-competency and self-worth, whereas internal attributions for failure tend to generate feelings of guilt. Although performers may feel grateful or thankful to others who help them to succeed, there also appears to be a danger that such attributions will cause persons to become more externally oriented. These external attributions could undermine performer's sense of personal control over performance outcomes. What would then be predicted to happen to motivation? It should decrease.

Weiner's Model of the Attributional Process

Although several attribution theorists have proposed models of the attibutional process, Weiner and colleagues have provided the most popular model. This model has two very important aspects; causal elements and causal dimensions.

Causal Dimensions

The four causal elements identified above can be classified along three dimensions locus of causality, stability, and controllability. Locus of casualty refers to the location of the cause in terms of being either internal or external. For example, ability is classified as being either internal, whereas task difficulty is classified as being external.

Stability pertains to whether or not the element is changeable. Effort is variable or changeable, whereas ability is ordinarily perceived to be stable. Controllability refers to whether or not the element is perceived to be under the control of the performer. For example, luck is an uncontrollable element, whereas effort is under the performer 's control. Controllability greatly affects motivation.

If, after failure, a performer attributes the outcome to a causal element not under that person's control then the performer does not perceive himself or herself to be the controlling factor in the outcome. If a performer attributes failure to lack of effort, he or she perceives a covariation between effort and outcome; if one element is changed or varied, then the other element should also be changed or varied. Thus the performer perceives himself or herself to be in control of the outcome of a response. This is a positive sign that achievement responses will recur.

ATTRIBUTION AND SPORT BEHAVIOUR

Achievement Motivation

Based on the theoretical framework of McClelland, Atkinson, and Lowell, a major area of research concerning individual differences has been achievement motivation. From this perspective, persons are perceived to have one of two primary motives, either to approach success or to avoid failure. "Approach-success" person are sometimes referred to as high achievers, whereas "avoid-failure" persons are often called low achievers.

Several researchers have demonstrated a consistent pattern concerning achievement motivation and causal attributions. High achievers tend to take personal responsibility for success by attributing it to their own ability. Low achievers deny personal responsibility for success by attributing causality to an external source. Thus high achievers reap affective benefits associated with success in achievement situations; low achievers do not. High achievers also see themselves as personally responsible for failure by attributing it to lack of effort.

Therefore they perceive that they can change their own destinies simply by varying the amount of effort exerted. Behaviourally, you should be able to identify high achievers by their motivation to enter into achievement settings such as sport. In addition, these are the performers who tend to persist at a task even after failure because they believe that if they just try harder, they will be successful. On the contrary, given the chance, low achievers will avoid such achievement situations.

Also, because they perceived that failure is caused by a lack of ability, low achievers do not tend to try again or persist other failure. They simply stop trying. They become passive. Of greatest interest in the coaching or teaching of sport skills is whether the maladaptive pattern of attributions exhibited by low achievers can be changed so that it mirrors that of high achievers.

Furthermore, if such change is possible, will it be reflected in behaviours such as approaching achievement settings and persisting after failure? If our thoughts and perceptions really to determine our behaviour, then a change in attributions should result in an observable change in behaviour. The procedure used to alter attributions for failure is called either an attribution retraining treatment or an achievement change program.

The goal of such programs is to alter the attributions for failure and therefore increase achievement striving or persistence. A good example of such a program is that used by Dweck. Dweck and her colleagues hypothesized that a long-term training program that taught learned helpless children to take personal responsibility for failure by attributing it to lack of effort would subsequently increase task persistence.

They used two experimental groups. One was a "success-only treatment" wherein the subjects were able to complete all experimental trails successfully. The success-only treatment was included because some theorists and many teachers believe that children can avoid the unpleasantness of failure if they do not have to experience it.

The other experimental approach was an "attribution retraining treatment". This differed from the success-only treatment in that failure occurred in about 20% of the experimental trails. When failure was evidenced, it was clearly attributed by the experimenter to lack of effort. The results of the study showed strong support for the assertion that if persons are taught to attribute failure to lack of effort then they will begin to show increased persistence behaviour after failure.

According to Dweck, failure can become a cue that effort must be increased. The use of the effort attributions indicates to the person that failure results from a lack of motivation rather than a lack of ability. Motivation is changeable and under the control of the performer. Also of interest was the finding that the subjects who experienced only success did not show any consistent improvement in their response after failure. On the contrary, performance was markedly impaired after failure. The notions of learned helplessness, low achievement motivation, and reattribution training have distinct possibilities for applications within sport situations.

Clearly, reatribution techniques could be used to increase the persistence behaviour of persons in sport settings. However, caution should be exercised. We suggest you follow the steps below when using reattribution procedures in sport and motor skills settings:

1. After the performance attempt, ask why either the successful or unsuccessful outcome occurred. Positively reinforce effort attributions.

2. Select a goal that can definitely be attained if the performer exerts sufficient effort. This selection process is of critical importance.

3. When the performer is consistently achieving about a 80% to 90% success rate, establish a new goal and repeat the sequence of steps.

4. Provide preperformance instructions that tell the person that successful performance depends on exerting effort.

These steps can help facilitate the perception that effort determine outcome. And because the performer really is achieving success, perception of his or her ability should be enhanced. Once both of these perceptions are held, the previously low achiever is beginning to think like a high achiever, and motivation to perform and persist at similar tasks in the future should increase. You should be able to observer these motivational changes in the form of increased approach and persistence responses. Dweck has also speculated on how aspects of learned helplessness might be manifested even in elite athletes.

In such cases, the helplessness would most likely not be observed over the long run. However, during a game or a match, if a failure begins to trigger thoughts of incompetence or pending doom, then feelings of helplessness will probably follow. Dweck uses the example of tennis star Evonne Goolagong-Cawley, who was fine when things were going her way; but when events didn't go in her favour, she tended to fall apart. The statements made by athletes relative to control and the attributions given for mistakes can provide valuable insights into destructive perceptions the performers may hold. That information can then be used to structure "mini" reattribution procedures. This should be particularly effective with high-level athletes whose ability level is not at issue.

Gender Differences and Sex-Role Stereotypes

Research studies concerning differences in attribution patterns as a function of gender have indicated that men and women generally follow the same patterns. However, there is evidence to suggest that sex-role strereotypes may exist. In the only investigation that has taken a developmental approach to the study of attribution within sport settings, Bird and Williams assessed observer ranging in age from 7 to 18 years. Their results indicated that as the youngsters got older, sex-role stereotypes for sport performance become more apparent.

The younger children attributed outcomes for both boys and girls similarly. However, by age 13, sport outcomes for girls were attributed to the external factor of luck, whereas those same outcomes for boys were ascribed to the internal factor of effort. Thus their findings indicated that by adolescence, boys are perceived as being in control of their own achievement outcomes; girls are viewed as being at the mercy of the environment. It is also important to point out that these sex-role steretypes were held equally by both boys and girls. However, other evidence suggests that when children are making judgements about their own performance, rather than the performance of others, this pattern does not always arise. For example, Iso-Ahola examined the sex-role stereotypes of fourth-grade boys and girls. He found no indication of differences in the attributions for own success or opponent's success as a function of gender.

The differences between the Bird and Williamns findings and those of Iso-Ahola indicate that attributions made by sport participants may not be identical to those made by observers. Interestingly, the same study uncovered a provocative stereotypes on the part of boys. When a boy lost to another boy, he would not accept lack of personal ability as the cause. However, when a boy lost to a girl, he would not accept his lack of ability as the cause. It appears that boys may become self-protective after losing to a girl. Such a loss goes against the sex-role stereotypes that boys are more competent than girls at sport skills. Taken as a whole, the research evidence provides a strong basis on which to argue that the cognitions that a person holds and the attributions a person uses to explain the causes underlying performance outcomes have definite effects on motivation. In addition, attributional patterns may vary as a function of achievement motivation, self-serving biases, and gender of the performer.

SELF-EFFICACY THEORY

Another major contribution to sport motivation research in Bandura's self-efficacy theory. According to Bandura, self-efficacy refers to the strength of a person's conviction that he or she can successfully make a response that is required to produce a certain outcome. Thus self-efficacy theory draws strongly on a cognitive interpretation of motivation. Bandura proposes that cognitive processes play a dominant role in both the acquisition and retention of new behaviour patters.

Information is acquired either directly or vicariously and is then coded symbolically into a cognitive representation that can guide future behaviour. In terms of motivation, the cognitive representation is thought to be capable of predicting future outcomes: "The capacity to represent future consequences in thought provides one cognitively based source of motivation. Through cognitive representation of future outcomes individuals can generate current motivators of behaviour". The cognitive representation can also influence motivation by functioning in the goal setting and self-evaluation processes. For example, an athlete might cognitively set a goal of running a race at a specific pace.

The goal setting serves as a motivator of behaviour. After the run, an evaluation is made against the standard represented in cognition. If success was the outcome, self-reward or self-reinforcement should follow. This is a pleasant experience and can motivate future behaviour. Failure should cause dissatisfaction and can influence the athlete to make behavioural changes that will lead to more successful performance. Thus the cognitive processes associated with both the expectancy of future behaviour and the evaluation of the outcome of that behaviour can affect motivation.

Efficacy Expectations versus Outcome Expectations

Bandura emphasized that a distinction must be made between efficacy expectations and outcome expectations. An efficacy represents the strength of the conviction that the person is capable of producing a certain outcome. This is why the box labelled efficacy expectations is located between the person and the behaviour.

From Bandura's perceptive the motivational force underlying behaviour is directly related to the nature of the efficacy expectation held by the performer. Thus his perspective on motivation is called self-efficiency theory. Now notice that the box labelled outcome expectations is situated between the behaviour and the outcome. This is because an outcome expectancy pertains to a person's estimation that a particular behaviour will produce a certain outcome.

Another way of looking at these two expectations is in terms of their informational versus motivational roles. The knowledge that a certain response is required to produce an outcome does not guarantee that the performer will attempt the response. Usually, a behavioural response will occur only if the performer hold the conviction that he or she can successfully enact the required response. For example, a beginning tennis player may know that the correct response for a job is a smash. However, the performer may not attempt that response because of the perception that he or she cannot successfully execute the required stroke. Efficacy expectations are viewed as being specific to given activities rather than being global, or generalized. A person's perception of self-efficacy can influence both choice of activities and persistence of coping efforts in the face of anxiety-producing situations. Perceived self-efficacy can also determine the amount of effort a person expends and how long he or she persists when faced with obstacles.

Self-Efficacy and Sport Behaviour

Participant Modelling

Participant modelling consists of three fundamental elements; modelling, guided participation, and the attainment of performance success. Fundamentally, it consists of structuring the environment with response-induction aids through the following procedures:

1. Once the learner is able to successfully execute the task requirements with the assistance of the model or other physically aids, then such support should be gradually withdrawn.

2. The "therapist" or another model performs the activity in easily mastered steps. The model should demonstrate effort, should be as similar as possible to the observer, and should demonstrate achievement through effort.

3. The participant modelling procedure is complete when the performer is able to master the task in the absence of any external assistance.

4. The performer attempts the task with the guided, physical assistance of the teacher or therapist. The goal is to achieve success.

Sources of Efficacy Expectations

Bandura had identified four sources of information that influence efficacy expectations:

1. performance accomplishments,
2. vicarious experience,
3. verbal persuasion, and
4. emotional arousal

Performance accomplishments are seen as the most powerful source of efficacy information because this information is based on a person's own personal mastery experiences. Multiple successes lead to strong positive efficacy expectations, whereas repeated failures generate strong negative efficacy expectations. Because performance-based information is so potent in terms of affecting perceived efficacy, performance based behaviour change programs are the most powerful.

Vicarious experiences provide people with a great deal of information and can influence the perception of their own efficacy at a task through the observation of models. For example, when a person observe someone else perform a threatening task without adverse consequences, this experience can trigger an expectation by the observer that if he or she intensifies effort and persistence, improvement or success can follow. When vicarious modelling is used to reduce observer threat, several meddling variables need to be considered, including the amount of effort displayed, the similarly of model to observer, and the clarity of the outcome conveyed.

Verbal persuasion is a common technique used to influence or change behaviour. Coaches frequently use verbal suggestions to convince athletes that they are indeed capable or mastering or coping successfully with a task. Neither verbal persuasion nor vicarious modelling is as powerful in affecting self-efficacy as the performance based approaches.

Emotional arousal is frequently used to provide information concerning a person's competency. According to Bandura, "People rely partly on their state of physiological arousal in judging their anxiety and vulnerability to stress". Very high states of arousal or anxiety can have negative effects on performance; therefore a person who experiences high levels of arousal or anxiety usually expects to perform poorly. From Bandura's perspective the relationship between arousal and self-efficacy is inverse; as arousal increases, self-efficacy decreases. Although there have been some critics of Bandura's intepretation of the role of arousal and anxiety, it would only complicate matters unnecessarily to present the conflicting arguments here. As mentioned previously performance-based procedures are the most powerful vehicles for behaviour change. These procedures are designed to provide a person with direct mastery experiences. Thus, from the perspective of self-efficacy theory, successful performance is the primarily vehicle for changes in perceived efficacy.

COMPETENCE AND INTRINSIC MOTIVATION

Several different theories have been proposed to explain our apparent motivation to feel competent and in personal control of the events that affect us. Although the names of the theories have varied, from instrinsic motivation to personal causation to effectance motivation and competence motivation all these perspectives views humans as being positively motivated to strive for accomplishment and achievement. A prime example of such striving is the desire to achieve excellence in sport.

Although any of the current theories is slightly different from the others, each alludes somewhat to three fundamental elements within the motivation process. First, people have a need to feel competent, self-determined, and in control. Second, the basic goal is to have an effect on the environment or to master the environment. Third, the reward for competency based or mastery responses is intrinsic. It is reflected in positive affect and self-reinforcement as opposed to the receipt of an extrinsic reward. Of the theoretical formulations that have been developed to describe and explain mastery behavior, one that appears well suited for application to the realm of sport is Harter's competence motivation.

Development of Competence Motivation and Instrinsic Motivation

The essence of the competence motivation position is that humans are born with the capacity and motivation to experienced joy or happiness generated by their own actions or responses. With increasing age, however, there are wide variations, or individual differences, in the degree to which people seek out mastery experiences such as sport or enjoy the positive affect that accompanies those experiences. Infants and very young children participate in mastery activities simply for instrinsic reasons, for the challenge and the resulting joy and fun. They do not need external rewards to motivate their behaviour; it is enough to fulfil the needs of curiosity, control, and challenge. It's fun! However, with increasing age mastery seeking declines for many persons. What causes this decline? Harter has identified the nature of the evaluative response of significant other's to the mastery seeking behaviour of a person as a crucial variable in that process.

Throughout this text we have stressed the potency of evaluative responses in affecting behaviour and the power of the observational learning process in regard to acquiring skills and knowledge. Both factors affect the development of mastery behaviour. When a child enacts a mastery-related response, parents and significant others, such as coaches, evaluate it.

The usual basis for this evaluation is the degree to success the child's response produced. Thus, based on their own standards of correctness or excellence, significant others convey powerful evaluative information to the youngster. The child uses that information to begin the internalization of his or her own system of mastery goals, criteria for judging future success or failure, and establishment of perceptions concerning control. It is also important to note that by evaluating the child's action on the basis of response outcome alone, the positive emotions the child experienced during the mastery response are often ignored.

This sends a clear message to the child that it is outcome, rather than positive affect, that is important, even though that is not the sole message that most adults would want to convey to children. Besides providing competency information on the basis of evaluating performance, significant others such as teachers and coaches influence the development of competence motivation by functioning as models. The child observes the standards used for evaluation by significant models and then internalizes this information in developing a sense of competence.

Sense of competence relates directly to a person's perception of his or her ability. The child also notices and can internalize aspects of the affective reaction of the model. Based on her work with Connell, Harter further proposes that as the internalization of mastery goals, criteria for success and failure, and perceptions regarding sources of control becomes more crystallized, the child develops a cognitive representation of perceived competency.

Once the cognitive representation is adequately developed it is capable of generating its own affective reactions to master-seeking responses. More specifically, up until about the age of 5 or 6 years, children's affective reactions appear to be undifferentiated. Events are either good or bad, happy or sad. By school age, and with the development of the internalized, cognitive representation of sense of competence, affective reactions to events or responses can be differentiated.

Reactions can be more specifically described, such as feelings of pride, happiness, or anger. When affective reactions are positive, they tend to enhance the probability that mastery striving will occur. Negative affective reactions tend to undermine mastery behaviour by reducing feelings of competency. Shame or anxiety about potential failure in sports should reduce the probability that a mastery-oriented response will recur. At about the same time that affective reactions become differentiated, competence is also differentiated into the cognitive, social, and physical domains.

When this occur, there are observable differences in the degree to which a person is motivated to perform in each of those domains. As we all know, not all highly skilled athletes are equally motivated on the playing field and in the classroom. Similarly, not all "eggheads" are also "jocks".

Extrinsic Rewards, Competency, Motivation, and Sport

Doubts about self-competence can make a person more externally oriented. Therefore by school age and continuing through later life there appears to be a strong connection between a person's cognitions relative to goal setting, criteria for success and failure, and perception of control and his or her sense of competence. This is true at all levels of sport, whether youth sports, interscholastic, intercollegiate, or professional sports.

According to cognitive evaluation theory, the performer's perception of personal competence and self-control over behavioural outcomes are extremely powerful variables underlying intrinsic motivation. When people take part in activities just for the sake of doing so rather than for extrinsic rewards, they are said to be intrinsically motivated.

Therefore, when a person feels that he or she is competent and is personally responsible for sport outcomes, that person is generally intrinsically motivated to participate. A major factor influencing the perception of personal responsibility for success and failure is the application of extrinsic rewards.

When intrinsically motivated activities are rewarded extrinsically, several possible consequences exist. From Deci's point of view, future intrinsic motivation can remain the same, be enhanced, or be depressed. His analysis is founded on the notion that rewards can serve two different functions: controlling and informational.

Controlling reares. If the controlling aspect is emphasized, persons begin to perceive a correlation between the enactment of the behaviour and the receipt of a reward. They begin to perceive that the reward is controlling their behaviour. Under such conditions intrinsic motivation should decrease. This is because a contingency has now been perceived between making a response and receiving a reward. Persons may then reason some way such as "no reward, no response." For example, an athlete who is rewarded for participating in an instrinsically interesting activity, such as playing softball or baseball, might decide not to play that game in the future if there is no extrinsic reward for doing so.

How many professional athletes would continue to participate in their sport with no monetary rewards? Probably not very many. Yet for the majority if certainly was not money that motivated them to participate initially. How many athletes on scholarship would be motivated to continue their athletic participation if the scholarship were no longer available? According to Ryan not very many would. He found that 75% of the athletes he assessed were primarily motivated by their scholarship rather than intrinsic factors.

International rewards. Informational rewards tell the performer something about his or her competence at a task. When rewards are given contingent on quality of performance outcome, they are powerful sources of information about competency. When the informational aspect of a reward is both salient and positive, then intrinsic motivation should be enhanced.

However, under those same conditions, if the nature of the extrinsic information is negative, then it can have a detrimental effect on intrinsic motivation. Keep in mind that the most frequent extrinsic reward that a coach gives is positive verbal feedback. How does positive feedback affect intrinsic motivation? Vallerand and Vallernad and Reid found that positive feedback enhanced both instrinc motivation and perceived competence, whereas negative feedback decreased intrinsic motivation and perceived competence.

Their results argue strongly that coaches should use positive rather than negative feedback whenever possible. Be positive? The analysis presented above argues that extrinsic reward should only be given when the quality of the performance meets some pre-established standard or excellence and when the rewards provide information relevant to competence and self-determination. The performer should be made aware of the basis on which the reward was provided. For example, before performance the coach and athlete might determine a specific performance goal that is related to one of the three dimensions of motivation. In volleyball an intensity-related goal could be to increase the number of blocks attempted by three.

If that goal is accomplished during the match, then the coach should verbally praise the player. Therefore it is not only the objective result of the blocks attempted that is important; so are the number of blocks attempted. In addition, the controlling aspect of rewards should be minimized or eliminated. This is done to avoid having the performer adopt a perspective that the activity is done to receive a reward because such a perceptive directs the performer's focus toward external rather than internal motivation. There are many occasions within sport settings when participant's perceptions and motivational orientation can be changed by introducing extrinsic rewards.

The giving of medals, trophies, and certificates is common is organized athletics. Is it possible that these awards can undermine athlete's sport motivation? Yes. However, several other factors should be taken into consideration. There is some research evidence that the nature of the reward affects whether it will or will nor decrease intrinsic motivation.

Deci and others have shown that verbal rewards or feedback do not usually have an adverse effect on intrinsic motivation. However, more tangible rewards, such as medals or money, can have a negative effect. Recall, however, that Deci also proposed that the nature of rewards must also be looked at in terms of their controllability and informational components. If by giving an athlete an extrinsic reward a perception develops that behaviour is being controlled externally, then this perceived change in the source of the reward will indeed reduce intrinsic motivation.

Under those same conditions, the performer's perception of competence and self-determination should also be affected adversely. For example, suppose young boy has tried out for a soccer team simply because he enjoyed that activity. Each night after practice he goes home and expresses all the fun and enjoyment he experienced that afternoon. His parents then decide that they will begin attending the practices and games; they get extremely involved and decide that their son could do even better and would have more enjoyment if he received a reward for his participation in soccer.

The parents tell him that they will take him out for ice cream every night after practice. They follow this procedures for a few weeks; however, with the passage of time their interest diminishes and other events conflict with soccer. The parents attend of fewer and fewer practices and games. As a result there are fewer and fewer trips to the ice cream parlor. What would be predicted to happen to their son's intrinsic interest in playing soccer? If a strong connection had been made between playing soccer and getting a reward of ice cream, the removal of the extrinsic reward should diminish interest in soccer.

If, on the other hand, extrinsic rewards are provided on the basis of the quality of performance, then they can be sources of information and can enhance intrinsic motivation. Such rewards can also increase perception of self-competency and self-determination. Suppose that at the end of the soccer a player received an award for being the most improved player on the team.

How should this reward affect her sense of competence and intrinsic motivation? It should indicate that the efforts put forth throughout the season have paid off. Her perception of self-competence should be enhanced and her intrinsic motivation should be increased. The misuse of extrinsic rewards in sport settings can have drastic and negative effects on motivation.

Most participants perceive sport activities as fun and enjoyable in and of themselves; the introduction of sophisticated systems designed to reward participation in such activities can be quite detrimental. Several scholars have suggested that, particularly in youth sports, parents and coaches should redirect attention away from the idea of providing extrinsic awards based on the outcome and on to the affective aspects of the game. Other theorists have suggested that rewards should be based on the particular performer's achievements as compared with past experience rather than on comparison made between or among the player.

Achievement Goals

Roberts had identified a constellation of three different types of achievement goals that might determine why people participate in sport activities:

1. to seek social approval,
2. to demonstrate ability and
3. to be task-involved.

According to Roberts, some performers participate in sport activities primarily to gain the approval of others such as teammates, spectators, and coaches. Game outcome is of less importance than is the social approval from significant others.

Such approval is usually contingent on demonstrating effort; therefore persons who have social approval as the predominant motivational force do exhibit effort but do so for external reasons. From Robert's perspective, persons who have a goal of competitive ability participate in sport to see how their ability matches up to that of other performers.

These persons consistently employ some comparison processes whereby they evaluate their own ability against that of their opponents. Such ability comparisons are most often made on the basis of competitive outcome because persons whose motivation is objective criteria such as winning the game or match. As a result, outcome become the most important criterion for sport participation. This focus on outcome as the primary criterion for evaluation can be detrimental to achievement strivings.

Suppose, for example, that the athlete really does try hard but constantly loses. What should be the resulting effect on perceived competence? It should decline. After all, the performer did exert effort, but lost anyway; therefore the explanation must be that ability is lacking. The sport mastery orientation is closely related to competence motivational and intrinsic motivation.

An athlete who has the predominant sport participation goal of performing well, regardless of outcome, is considered to be sport-mastery oriented. This type of performer cares about improving skill rather than beating others. Levels of ability or achievement is judged in comparison to that performer's past achievements rather than by making comparisons against others. In terms of competence, self-determination, and positive affect, it seems more desirable that people participate in sport activities for reasons of sport mastery as opposed to competitive ability or social approval.

One way to enhance adoption of a sport-mastery orientation is to use a reinforcement system that provides rewards contingent only on improvement in an athlete's performance compared with that person's previous performance. On the contrary, adoption of an outcome-oriented reinforcement contingency can underline perceived competence. Furthermore, because only one person or one team can be the winner, all the others must be losers.

If outcome is the only criterion used to evaluate

sport behaviour, then all these losers will experience negative affect as a result of sport participation. It seems much more desirable to use a strategy in which individual efforts and improvement are rewarded regardless of outcome. Another important aspect of Roberts' work involves the question of how sport achievement orientation may relate to dropping out of sport.

Of the three orientations identified, social approval, competitive ability, and sport mastery, which do you think would most relate to children dropping out of sport? If you thought the answer would be the social approval orientation, think again. Such people can very easily gain social approval in sport contexts because if they exhibit effort in the sport situation, then significant others will probably reward them. Now think about the difference between being focused on competitive ability and being sport mastery-oriented.

In the case of competitive ability, evaluation of ability is generally made on the basis of the game outcome. Failure indicates lack of ability regardless of individual effort. One way to avoid such failure evaluations is to simply drop out of sports. However, Roberts points out that this does not usually occur until around age 12. It seems that before that age children do not have sufficient cognitive development to distinguish between effort and ability.

At about age 12, children realize that ability, to a large extent, determine sport outcome. They begin to understand that for some children even large amounts of effort are not going to ensure success. In addition, most significant other emphasize success as being based on outcome. Thus it seems reasonable that if an adolescent perceives his or her ability to be low, then dropping out of sports is one way to avoid the negative consequences associated with failure. This explanation may also account somewhat for the significant numbers of women who drop out of sports during their teenage years and avoid public display of their sport ability throughout later life.

The recent interest and participation of some women in noncompetitive activities such as jogging and aerobics also lends credence to that explanation. It was interesting to observe the marketing strategies used in many 'women' magazines during the beginning of the jogging craze.

One such magazine promoted the sale of jump ropes to women by saying they could work out in the privacy of their own homes, where no one could see them! This phenomenon is also evident in many health clubs that have "women only' exercise rooms. Roberts has also suggested than an alternative to dropping out would be to redirect achievement orientation to the task itself.

The goal should be sport mastery, and emphasis should be placed on performance of the activity rather than outcome. Social comparisons should be avoided and the use of self-evaluations should be encouraged. If you want to enhance individual sport motivation, then the selection, intensity, and persistence dimensions of motivation, not outcome, should be used as criteria for self evaluation and positive affect.

5

ANXIETY, STRESS AND AROUSAL

The story is told of an African hunter who, after losing his weapon, was pursued by a lion. He was running along as fast as he could go with the lion in hot pursuit, when he spotted a tree limb about 12 feet off the ground. Without breaking stride, he leaped with all his might, hoping to jump higher than he had ever imagined possible. As luck would have it, he missed the limb going up, but he caught it coming down!

This tale illustrates an interesting fact bout the phenomenon of arousal. When extremely aroused, we are often capable of astonishing feats. However, not that the man in the story missed the tree limb going up. Luckily for him he caught it on the way down–but what does this say about his accuracy?

The state of being anxious at a result of an environmental stimulus is associated with an elevation in arousal. For this reason, many people have assumed that the two states are the same. When confronted with physical violence, for example, an individual will typically feel a rush of fear and anxiety that is associated with an increase in such things as heart rate and breathing. An increase in arousal, however, can also be associated with the emotions of joy, happiness, and exhilaration, which often accompany the thrill of sports victory. Arousal is a neutral physiological phenomenon that can be associated with both negative and positive affect, yet is not synonymous with either.

NEUROPHYSIOLOGY OF AROUSAL

Defining Anxiety and Stress

The following story about a found athlete illustrates the potential debilitating effects of anxiety on athletic performance. Ryan is a physically gifted sixteen-year & old athlete.

He participates in several sports for his high school during the academic year and plays summer baseball. However, his favourite sport is track and field, which is primarily an individual sport.

Ryan is a highly anxious young man with a tendency toward perfectionism. In Ryan's particular case, these traits had very little negative effect on his performance in the team sports he played.

He would often get uptight about a big game, but he could always rely upon his team-mates ton help him out. The fact that team games involved other players seemed to help control the negative impact his anxiety could have had on his performance. Ryan occasionally "clutched" during baseball games, but the outcome of the game was rarely affected. Usually, only Ryan and Ryan's parents were aware of the anxiety and tension that was boiling within.

However, Track and field was a different matter. Ryan was a sprinter and hurdler. His physical power and mesomorphic build made him especially well equipped for running and jumping events that required speed and leg power. Unfortunately, his basic anxiety and worry about failing had serious effect on his performance during competition. During practice, Ryan always did well.

In fact, during the years of high school Ryan had never lost a race to a team-mate during practice. In actual competition, things were different. Yan began preparing mentally for his races days in advance of the actual competition. During the days and hours preceding competition, his anxiety would rise to fearful levels. By the time actual competition came, Ryan could hardly walk, let alone run or jump. Several times he had to vomit before important races.

His coach talked to him a great deal about learning to relax and not worry about the race, but didn't give him specific suggestions on how to accomplish this. Finally, the coach decided to remove Ryan from his favourite events because he was actually a determent to the team.

This was more than Ryan could take. He approached the coach one day and announced that he was going to give up athletics altogether and concentrate on his studies. This story has successful conclusion, but it will be shared later, at the beginning of the chapter on arousal adjustment strategies.

We can thing of anxiety as being situation-specific anxiety response to a threatening stimulus is referred to as state anxiety. State anxiety is an immediate emotional state that is characterized by apprehension, fear, tension, and an increase in physiological arousal.

Trait anxiety, on the other hand, is a personality predisposition. It is a predisposition to receive certain environmental situations as threatening, and to respond to these situations with increased state anxiety. According to Endle, and combination of five factors include

(i) threat to a person's ego, (iii) Threat of personal harm, (iii) ambiguity, (iv) disruption of routine, and, (v) Threat of a negative social evaluation.

Anxiety has often been linked with the term stress, yet stress is a much broader and more comprehensive term. In the classic sense, stress is the "nonspecific response of the body to any demand made upon it". Stores, like arousal, is a neutral physiological response to some sort of stressor.

The stressor could be in the form if physical exercise, joyful excitement, or threat to the body. All instances of stress, regardless of their origins, result in identical in the body. As explained by Hans Style, there are four basic variations of stress.

Hyperstress occurs when the amount and degree of stress exceeds our ability to the stress. Conversely, hypostress occurs when we suffer from a lack of sensory stimulation. Eustress is conceptualized as "good stress," and is manifested in the form of joy, exhilaration, and happiness. Conversely, distress, or "bad stress, " is manifested in the form of tension, anxiety, and worry.

In order to neutralize the negative effects of stress in our lives, we must strike a balance between too much and too little stress, find as much eustress as possible, and do al. in our power to minimize or eliminate the debilitating effects of distress. Based on their respective definitions, state anxiety and distress are virtually identical.

Another interesting and useful way to conceptualize stress is in the form of process. The stress process, as described by Martens and by McGrath, refers to the process associated with the manifestation of state anxiety.

This process is explained in terms of an objective demand, a perceived threat, and a state anxiety reaction. The objective demand represents the situation that the athlete is placed in; it is considered a stimulus. Whether or not the athlete sees the objective situation as threatening depends on her subjective evaluation of the situation and the interactive effect of trait anxiety.

The relationship between the objective situation and the response can also be explained in terms of an imbalance. Distress will occur if the perceived demand i not balanced by the athlete's perception of her ability to respond effectively to the threat.

Two researchers were studying the effects of fear of drawing on the physiological responses of a subject. The subject was strapped to the side of a swimming tank with the water steady rising. For some reason the researchers left the test area and forgot about their subject.

When they remembered, they were aghast and numb with fear. Dropping everything. They raced to the test area to find the water level dangerously high. Quickly, they unstrapped the subject and pulled him from the water. Safely on the pool deck, they asked the subject if he was frightened.

The subject responded that he wasn't at all worried, because it was just an experiment and he knew that the researchers wouldn't let any harm come to him! The subject perceived the test situation to be non threatening, and therefore the state anxiety reaction was not evoked.

Neurophysiology of Arousal

To understand arousal is to understand what basic changes take place in the body when the athlete is activate. Arousal is the degree of activation of the organs and mechanisms that are under the control of the body's autonomic nervous system.

In a subsequent chapter of this book, various strategies for adjusting the level of arousal in the athlete will be discussed. For this reason, it is critical that the sport psychologist understand the neurophysiology of arousal.

Brain Mechanisms

Fundamental to the concept of arousal is the notion that levels of activation come under control of progressively higher levels of the nervous system. Structures of the central nervous system that are closely related to the phenomenon of arousal include the cortex of the brain, the hypothalamus, and the articular formation.

The cerebral cortex is the area of the brain responsible for higher brain functions and conscious thought processes. The functional part of the cerebral cortex is comprised of a thin layer of neurons. The Total cerebral cortex contains approximately ten billion neurons. The electrical activity of the cartes is measured with the electroencephalograph.

The EEG can monitor frequency and amplitude of

electrical potential changes in the brain. High states of arousal are associated with EEG waves that are desynchronized, fast, and of low amplitude. Low states of arousal are associated with a synchronous pattern of EEG waves.

The Hypothalamus, a part of the midbrain, has been shown to be an important part of the arousal system. Lesions of the posterior section of the hypothelamus cause sleep and drowsiness, while electrical stimulation of these same areas has ben reported to cause alertness and excitement.

Stimulation of the posterior hypothalamus also causes the secretion and release of catecholamine by the adrenal medulla.

The reticular formation is a complex set of neurons and nuclei that extends throughout the brain stem from the medulla to the posterior hypothalamus. This network of neurons sends diffuse fibres throughout the nervous system and cortex. It appears that the reticular formation organizes sensorimotor behaviour through its interconnections with the cortex, hypothalamus. And nervous system.

Along with the cortex and hypothalamus. The activity of the ascending reticular activating system is closely associated with the onset of arousal. The ascending axons of the ARAS facilitate the higher brain centre neurons. Stimulation of the ARAS of sleeping animals results in EEG waves resembling those they would have if awake. Lesions of the ARAS in cats result in somnolence. When an individual perceives a situation to be threatening or arousing, the articular formation is activated, and this in turn governs the arousal sequence.

Any type of sensory stimulus from the environment or from the cerebral cortex can initiate the arousal reaction at the level of the reticular formation. The reticular activating system responds very quickly to bring about a general activation of other brain structures. Almost simultaneously, the sympathetic nervous system is activated by centres in the cortex, spinal cord, brain stem, and hypothalamus. Stimulation of sympathetic nerves to the noradrenalin to be released into the circulation bloodstream. Until these drugs are re-absorbed into body tissues, the organism will remain in an aroused state.

Autonomic Nervous System

The nervous system in humans contains two major divisions: the peripheral nervous system, or the nerves in the skeletal muscles of the body, and the autonomic system or the nerves in the smooth muscles and glands of the body.

That part of the nervous system directly related to activation and arousal is the autonomic nervous system. It is autonomic in the sense that we do not normally have voluntary control over the organs and glands innervated by it. This is not entirely true, of course, but normally we do not control such bodily functions as heart rate, blood pressure, and respiration.

The autonomic nervous system is itself divided into two divisions, the sympathetic and the parasympatholytic nervous systems. The sympathetic division is primarily responsible for changes in bodily functions associated with arousal.

For example, it is the sympathetic division that brings about sweating of the hands, increased heart rate, pupil dilation, increased respiration, release of glucose from the liver, and decreased kidney output. The sympathetic division release catecholamines at the postganglionic innervation site of the gland or smooth muscle.

The sympathetic division tends to result in arousal of the organism, while the parasympathetic division selectively reduces the effects of the sympathetic division. Stimulation of the parasympathetic division results peripherally in pupil constriction, decrease in heart rate, decrease in respiration, and in general, a return to a haemostatic balance of bodily functions.

The sympathetic nervous system responds very quickly to environmental or cognitive stimuli, while the parasympathetic nervous system is comparatively slow. When confronted with a sudden physical threat, the body's physiological response to it is nearly instantaneous.

Preferred State of Arousal

Arousal varies along a continuum from deep sleep to extreme excitement. Yet researchers have consistently observed that organisms seek to attain a level of arousal that is ideal for the task at hand, that is, a level of optimal arousal.

Schultz called this internal drive for seeking a state of optimal arousal censoriousness. Duffy observed that the optimal degree of activation appears to be moderate– not too much arousal and not too little. The hypothesized curve representing the relationship between arousal and the quality of performance is quadratic, taking the form of an inverted U. Further discussion on its curve appears later in this chapter.

Electro Physiological Indicators of Arousal

Many electro-physiological methods exist for measuring arousal level. Some of the more common methods will be outlined here. We should remember that no single measure of arousal can be considered completely accurate; correlations among the various measures of arousal are very low. This means, for example, that someone exhibiting a fast heart rate might not exhibit decrease in palmar skin resistance.

Electro-cortical Activity As discussed earlier, the electrical activity of the brain can be measured by an electroencephalograph, which measures the amount of electrical activity put out between two scalp electrodes. An electroencephalogram is the tracing of the brain waves made by an electroencephalograph. Three types of brain waves are associated with the arousal states of sleep, awakeness, and excitement. Depression of EEG alpha-wave activity is considered to be a strong indicator of heightened arousal.

Biochemical Indicators The arousal, or activation, response in the brain triggers the release of catecholamines into the bloodstream by the adrenal medulla. Thus, adrenaline and noradrenalin in the bloodstream. The accurate assessment of catecholamines requires immediate blood samples, since they readily diffuse into body tissues.

Heart Rate The heart of an aroused person is easily obtained with an electrocardiograph, a device for measuring the electrical activity of the heart.

The heart rate can also be indirectly measured with a use monitor or by finger palpation. The heart rate is not considered to be a good single indicator of arousal. Its correlation with other more viable indicators is quite low.

Muscles Tension The electrical potential of muscles can be measured with an electromyograph, a device that measures electrical activity in a muscle. Woodworth and Schlosberg have shown that muscle tension levels are roughly equivalent to levels of arousal. Weinberg and Weinberg and Hunt have successfully used EMG recordings to indicate arousal levels in motor performance research.

Respiration rate of respiration is not entirely under the control of the autonomic nervous system, but it is still a fairly reliable indicator of heightened arousal. With a spirometer, a person's respiratory rate, tidal volume, inspiratory reserve volume, expiratory reserve volume, inspiratory capacity, and vital capacity can be measured.

Blood Pressure Arterial blood pressure can be measured with a sphygmomanometer. Blood pressure is an indication of the relative dilation or constriction of the blood vessels associated with the autonomic nervous system. Since blood pressure is generally monitored through repeated application of the pressure cuff and stethoscope, only intermittent recordings can be obtained. For this reason, blood pressure is not considered to be a very good measure of arousal.

Palmar Sweating In a threatening situation, activation are associated with an increase of wet from sweat glands on the hands. According to Harrison and

Mac Kinnon, sweat glands of the human palm do not function in response to environmental changes, but are activated by alerting stimuli.

Techniques for counting palmar sweat glands are detailed by Sutarman and Thompson. Johnson and Dabbs, and Dabbs, Hohnson, and Leventhal.

Galvanic Skin Response Associated with increased palmar sweating during periods of heightened arousal is a corresponding change in the resistance of the skin to the passages of an decrease in skin resistance, or galvanic skin response. A decrease in skin resistance to the passage of electricity from one electrode to another is equal to an increase in skin conductivity.

The galvanic skin response is mediated through the sympathetic cholinergic nerve supply to the skin response is mediated through the sympathetic cholinergic nerve supply to the skin and is attributed to changes in the number of active sweat glands. Therefore, increased skin conductivity is directly related to an increase in the number of active palmar sweat glands.

Multidimensional Nature of Anxiety

As indicated earlier in this chapter, Endler identified five different dimensions, or facets, of anxiety. Studies by Fisher and Zwart and Gould, Horn, and Spreeman have provided support for Endler's multidimensional theory of anxiety in the sport domain. Any psychological test or theory that focuses on only one dimension of anxiety at the expense of others is considered to be undimensional in nature. As indicated by Endler, Spielberger's trait and state inventories are somewhat unvarying in nature, because they do not address potentially important causes of anxiety.

Multidimensional anxiety theory has resulted in a number of inventories that approach anxiety from a multidimensional as opposed to undimensional perspective. A case in point is the well-known and often-utilized competitive State Anxiety Inventory-2 developed by martens, Vealey, and Burton. Relative to competition, the CSAI-2 assesses sport-specific cognitive and somatic state anxiety. Cognitive state anxiety is the mental component of state anxiety cased by such things as fear of negative social evaluation, fear of failure, and loss of self-esteem.

Somatic state anxiety is the physical component of anxiety and reflects the perception of such physiological responses as increased heart rate, respiration, and muscular tension.

Time-to- Event Nature of Pre-competitive Anxiety

Our ability to obtain independent measures of cognitive and somatic state anxiety has greatly enhanced our knowledge about the athletic situation. One of the factors that is believed to significantly influence the quality of the athletic experience is the level of state anxiety during the time leading up to competition.

This is referred to as pre-competitive anxiety. We now know quite a bit about the temporal changes in anxiety during the period of time leading up to and immediately following the beginning of the event. Pre-competitive cognitive anxiety starts relative high and remains high and stable as the time-to-event approaches. Conversely, somatic anxiety remains relatively low until approximately 24 hours before the event, and then increases rapidly as the even approaches.

Once performance begins, somatic anxiety dissipates rapidly, whereas cognitive state anxiety fluctuates throughout the contest as the probability of success/failure changes.

Differential Effect of Somatic and Cognitive Anxiety on Performance

The capability to ascertain the independent measures of cognitive and somatic state anxiety has increased our knowledge of the relationship between pre-competitive anxiety and athletic performance.

The nature of the relationship between competitive state anxiety and performance depends in part upon the dimension of state anxiety measured. For example, the relationship between somatic anxiety and performance tends to be quadratic in nature and takes the form of the inverted U. Conversely, the relationship between cognitive state anxiety and performance is linear and negative. As cognitive state anxiety increases, athletic performance decrease.

This is consistent with the volleyball findings of Cox When a un-dimensional measurement of state anxiety and performance is quadratic and takes the form of the inverted U. As somatic state anxiety increases, performance also increases up to an optimal level, and then decreases as somatic state anxiety continues to increase.

It would appear that the CSAI-2 has led to a major breakthrough in our understanding of the relationship between state anxiety and athletic performance. Worry and apprehension associated with cognitive state anxiety is a major hindrance to athletic performance.

The lower the level of cognitive state anxiety is, the better the athlete will perform. The notion that a "little" anxiety is good for the athlete is not true, if the anxiety is cognitive.

Conversely, increases in the somatic indicators of state anxiety are associated with increased athletic performance up to an uncertain optimal level, and then thereafter result in a decrement in performance. This is the classic inverted-U relationship between anxiety/arousal and performance.

Intensity and Direction Issues

In subsequent sections of this important chapter, I will introduce other factors that enter into this relationship. Among them is the notion of individual differences. The relationships. Among them is the notion of individual differences.

Any single individual or athlete might exhibit a pattern very different from that of the group. The CSAI-2 measure intensity of state anxiety and says nothing about the individual's perception of whether his response, be it high or low, would hinder or facilitate performance. The athlete's perception of whether his "felt" response, relative to a specific item on the CSAI-2, was good or bad for future performance has been labelled direction of state anxiety. Direction of state anxiety is the athlete's interpretation of whether his lift level of intensity was debilitative or facilitative relative to performance.

To take into account the directional component of competitive state anxiety composed of 27 items, each set to a 4-point Likert Scale running from "not at all" to "very much so." The revised CSAI-2 requires a second response from the subject relative to each item.

The second response requests that the individual indicate on a 7-point scale whether he thought the level of intensity he marked would be debilitative or facilitative. The actual scale runs from a -3 to a +3, with a minus indicating debilitative and a plus indicating facilitative.

Jones et al. and Jones and Swain administered the revised CSAI-2 to athletes and then categorized them as either being elite or non-elite. Based on the intensity measure of state anxiety, they were unable to differentiae between the two group, but based upon the direction component, they could differentiate.

The elite athletes viewed their intensity responses to be much more facilitative than the non-elite athletes thought their responses were. In a related investigation, Jones and Hanton divided athletes up into two groups, with high and low expectations about goal attainment.

The athletes who had a high expectation that they would achieve their goals were categorized in the high-expectation that they would achieve their goals were categorized in the high-expectancy group. Conversely, the athletes who had a low expectation that they would achieve their goals were categorized in the low-expectancy group.

Based upon the results of the revised CSAI-2 administration, researchers were not able to differentiate between the two groups based on anxiety intensity score; however, they were able to differentiate based on directional scores.

Individuals in the high-expectancy group viewed their intensity responses to be much more facilitative than those in the low-expectancy group considered theirs to be. Jones and Hanton interpreted these results to mean that athletes who expect to attain their goals feel "in control," and view anxiety symptoms as facilitative as opposed to debilitative. Based upon this new research, it seems clear that an athlete's perception of whether or not anxiety symptoms are facilitative or debilitative is more important than whether the absolute intensity of anxiety is high or low.

Shortened Versions of the CSAI-2

In recent years, the most common inventory used for measuring pre-competitive competitive state anxiety has been the SSA(-2. The inventory is 27 items in length, takes about 5 minutes for an individual to complete, and is administered as close in time to the start of competition as possible.

This generally ranges from 60 minutes before competition, for high-profile events, to 15 minutes before intramural events. Admittedly, a great deal of variation in pre-competitive state anxiety could occur during this time frame.

In addition, athletes are often reluctant to allow a sort of distraction during the time they are preparing mentally to allow any sort of distraction during the time they are preparing mentally for competition. With these reasons in mind, Murphy, Greenspan, Jowdy, and

Tammen developed a 3-item version of the CSAI-2 called the Mental Readiness Form. Improved version of the MRF were developed by Krane.

The two improved versions of the MRF were set to 11-item Likert scales and titled the MRF-Likert and MRF-3. In addition to the MRF, Cox, Russel, and Robb developed the Anxiety Rating Scale, which is also a 3-item version of the CSAI-2 set to a 7-item Likert Scale. Bot of these scales provided the researcher and practitioner with a non-intrusive way to measure competitive state anxiety prior to, or even during, competition. Neither Krane nor Cox et al. recommend the use of the shorter versions of the CSAI-2C over the 27-item parent test if time is not an important consideration.

In addition to the MRF and ARS described above, Stadulis, Didson, and Mac Cracken reported the development of a 15-item version of the CSAI-2 called the competitive State Anxiety Inxiet Inventory-2 for children.

This CSAI-2C is a shortened version of the parent test modified to include language appropriate for children ages ten to twelve.

The Sport Grid

Russell, Weiss, and Mendelsohn proposed the use of the Affect Grid as a single-item scale for assessing affect along the dimensions of high/low arousal and positive/negative affect. The basic concepts developed by Russell et al. were modified by Readeke and Stein into the Sport Grid.

The Sport Grid provides a simple and direct way for

an athlete to indicate with a single mark on a grid whether he feels physiologically aroused and whether he is experiencing positive or negative thoughts and feelings at a specific point in time.

Like the MRF and the ARS, described in the previous section, the Sport Grid offers a quick and non-intrusive way for an athlete to indicate how he is feeling immediately before an important competitive event.

Measurement of Anxiety

In recent years, the preferred method of measuring trait and state anxiety has been through the use of pencil-and-paper inventories. For a more comprehensive description and listing of anxiety inventories, see Ostrwo.

While pencil-and-paper inventories are the most common measures of anxiety, behavioural and physiological assessment can be very effective. One category of behavioural measurement is direct observation, where the experimenter looks for objective signs of arousal in the subject and records them. Such things in respiration could all be interpreted as behavioural signs of activation. Such a system was developed and used by Lowe for ascertaining arousal through "on-deck activity" of batters in Little League baseball.

Along these lines, Harris and Harries prepared a list of overt behavioural responses that can be used by the athlete to identify indicators of distress, or state anxiety.

From the perspective of applied field research, pencil and paper inventories and behavioural assessment techniques seem most feasible. This will remain true as long as electro-physiological indicators require cumbersome and expensive instruments. Advances in the field of applied psychophysiology, however, may change this perspective.

Improvements in the use of telemetry may soon make it possible to monitor an athlete's heart rate, blood pressure, and muscular tension while she is competing in such dynamic activities as swimming, batting in baseball, and sprinting in track. Landers has long argued that both physiological and psychological assessments of anxiety should be taken to measure anxiety and arousal. In this regard, however, it is important to point out that the relationship between physiological and psychological measures of state anxiety are quite low. Consequently, if both physiological and psychological measures of state anxiety are recorded simultaneously, it is possible that conflicting results may be obtained.

Relationship Between Arousal and Athletic Performance

Throughout this chapter, an effort has been made to avoid confusing the terms anxiety and arousal. In this section, however, it will be necessary to use the term arousal as somewhat synonymous with state anxiety.

This is the case because researchers have routinely employed a test of state anxiety as the primary means for determining a subject's arousal level. Consequently, most of the reported research will relate negative anxiety to sport and motor performance.

The primary focus of this section will be upon two main theories that purport to explain the relationship between arousal and athletic performance: inverted-U theory and drive theory.

Inverted-U theory includes many sub-theories that explain why the relationship between arousal and performance is curvilinear as opposed to linear in nature. Conversely, drive theory proposes a linear opposed to linear in nature. Conversely, drive theory proposes a linear relationship between arousal and performance. In the most elementary case, the distinguishing characteristics of inverted-U and drive Theory.

Inverted-U Theory

The inverted-U theory has been around for as long as the arousal/performance relationship has been studied. It simply states that the relationship between performance and arousal is curvilinear as opposed to linear, and takes the form of an inverted-U. While it is described as a theory or hypothesis, researchers such as Duffy and Malmo consider it to be an observed fact.

One of the difficulties encountered in testing the inverted-U theory with humans is our inability to precisely measure arousal. For Example, if in a particular study researchers fail to demonstrate that heightened arousal causes a decrement in performance, it is no particularly damaging to the theory.

The reason for this is that it can always be argued that for particular task, arousal was not enough. If it had been higher, we may argue, performance would have declined. The problem is that form a human rights standpoint, the amount of arousal researchers can induce is limited. For example, if arousal is induced through electrical shock, how much can the researcher elevate the voltage without violating the subject's rights? Not very much.

The foundation for inverted-U theory is the classic work of Yerkes and Dodson. Using dancing mice as subjects, Yerkes and Dodson set out to discover the relationship between arousal and task difficulty in their effect on performance. Performance was measured as the number of trials needed for the mice to select the brighter of two compartment shock. Task difficulty was manipulated in therms of the differences in brightness between two compartments. Results showed that the amount of practice needed to learn the discrimination task increased as the difference in brightness between the two compartments diminished.

These findings led to the Yerkes-Dodson law, which states, "an easily acquired habit, that is, one which does not demand difficult sense discrimination or complex associations, may readily be formed under strong stimulation, whereas a difficult habit may be acquired readily only under relatively weak stimulation".

Additionally, an optimal level of arousal is indicated for each task, Before and after the optimal point, performance drops off. This is the inverted-U.

Conversely, a lower level of arousal is best for a fine motor task such as putting in golf. Each sport skill has its theoretical optimal level of arousal for best performance. Regardless of which type of sill is being performed, They all conform to the inverted-U principle. Specifically, performance is lowest when arousal is very high or very low, and highest when arousal is moderate, or optimum.

Another important consideration relating to the Yerkes-Dodson law is skill level. Just as putting in golf is a complex activity compared to weight lifting, learning to dribble a basketball is more difficult for a beginner than performing the same task as an expert.

The optimal level of arousal for a beginner should be considerably lower than the optimal level for an expert performing the same task.

Evidence of an inverted-U relationship between athletics performance and arousal is documented in the literature. Klavora and Sonstroem and Bernardo were able to demonstrate that basketball performance is related to level of arousal, with best performance occurring at moderate levels of arousal and poorest performance at high or low levels. Similarly, Gould, Petlichkoff, Simons, and Vevera and Burton reported that best performance in pistol shooting and swimming, respectively, were related to somatic anxiety in a way consistent with inverted-U predictions. In a related study, Beuter and Duda observed that heightened arousal has a detrimental effect on the motor performance of children. Under conditions of optimal arousal, children perform smooth and automatic movement patterns.

However, under conditions of high arousal, movement pattern come under volitional control and are observed to be less smooth and efficient. The inverted-U relationship between performance and arousal has also ben documented with non-athletic tasks such as reaction time, auditory tracking, and hand steadiness.

The basic premise of cue utilization or attentional

narrowing theory is that as arousal increases. Attention narrows. The narrowing of attention results in some cause being dated out, first irrelevant cues and later relevant cues. When arousal is low, the attentional band is wide and both irrelevant and relevant cues are available. The presence of the irrelevant cues is distracting and causes a decrement in performance. At a moderate, or optimal, level of arousal, only the irrelevant cues are eliminated, and therefore performance is high. Finally, when arousal is high, attentional focus is narrow and both relevant and irrelevant cues are dated out. This results in a decrement in performance, as predicted by the inverted-U theory.

Cue utilization theory also addresses the problem of task complexity and learning. With a complex or unlearned motor skill, there is a greater number of task-relevant cues to manage. Consequently, with increased arousal, the probability of errors increases at a faster rate than it would for a simple motor skill .

The phenomenon of attentional narrowing is easily applied to a spot setting. When a football quarterback drops back for a pass, an optimal level of attention will cause a dating out of irrelevant cues.

However, if arousal becomes too high, the quarterback may either suffer from distractibility or gate out relevant cues as a result of his narrow band of attention. Research supporting the concept of attentional narrowing is strong, but has taken place primarily in a laboratory setting.

An exception to this general rule is a sport-related study reported by landers, Qi, and Courtet.

Taken as a whole, this brief review of attentional narrowing research provides support for Easterbrook's cue utilization theory. Increased arousal causes attention to become narrower and more restricted. This can enhance performance up to a certain point, but thereafter it causes a performance decrement.

Thus, attentional narrowing theory supports an inverted-U relationship between arousal and performance.

Signal Detection Theory

Another theory that predicts a quadratic relationship between arousal and performance is signal defection theory. Signal detection theory has not been field tested for the inverted-U concept; however, from a theoretical point of view it should be of interest to the reader.

In its simplest form, signal detection theory holds that the intensity of noise in the nervous system falls along a continuum ranging from low to high.

The addition of a signal to the noise naturally increase the neural activity. It is the subject's task to discriminate between noise alone and signal plus noise. The signal is typically a sound.

Repeated random presentations of test intervals in which a signal may or may not be presented is theoretically represented by two bell-shaped curves. One curve represents the noise distribution, and the other represents the signal plus noise distribution.

These two curve are believed to be normally distributed, with the greatest frequency of observation occurring at about the man of each distribution.

The task of participants in an SDT experiment is to respond by saying "yes" if they detect a signal and "no" if they do not. Obviously, the constant presence of neural noise makes the task potentially difficult. If the neural activity seems high, the participant will be likely to respond "yes." If the neural activity associated with a test interval seems low, when will be likely to respond "no."

A participant's ability to discriminate between the noise and the signal plus noise distributions is called the participant's "sensitivity." It represents the distance between the means of the N and SN distribution.

One important factor that determines whether a participant will give a yes or no response to a particular stimulus is the participant's response bias, or response criterion. Some people exhibit a very stringent criterion and refuse to responds "yes" unless they are positive they have heard a signal.

Others have a lenient criterion and will respond "yes" even if the signal appears weak. Participants with a stringent criterion miss detecting many signal, while those with a lenient criterion identify signals that do not actually occur.

Welford has hypothesized that a high ora low level of arousal has the effect of altering the position of a participant's response criterion. Increased activation of brain cells would increase the neural activity of any particular point along the division axis .

As this occurred, both distributions would tend to shift to the right due to increased neural activity. Since the response criterion is relatively stationary, shifting the relation to the two distribution. This process would be reversed during sleep or low arousal.

In the low-arousal situation, the errors tend to be a failure to detect a signal. In the high-arousal situation. The errors tend to be false. In the optimal-arousal situation, the errors are ideally balanced between false alarms and misses.

This model is very appealing, because one would expect errors to increase with high and low levels of arousal.

From this explanation, it should be clear that with extreme shifts in the response criterion form the left to the right, we have the inverted-U relationship between signal detectability and arousal. For example, consider what would happen in the high-arousal. For example, consider what would happen in the high-arousal situation if beta shifted clearly to the lift of both distributions.

This would result in a totally lenient response criterion, by which subjects would respond "yes" every time they were asked if they detected a signal. Assuming a signal was presented 50 per cent of the time, this would yield a 50 percent error rate. Conversely, if in the low-arousal situation, the criterion shifted completely to the right of both distribution, the subject would respond "no" to every lest observation. Again, a 50 percent error rate would result.

Thus, high and low levels of arousal yields the best performance.

In sport, the SDT model can be applied to officials who must routinely make split-second decisions. Was the pitch a ball or a strike? Was the runner safe or out? The SDT model is easily applied to these situations by simply assigning the two curves to each of two possible decisions.

The theory can be used be used to determine an official's bias and sensitivity. As with athletes, when the official is under aroused or over aroused, he will make more errors than when he is optimally aroused.

6

EMOTIONS, SOCIAL INTERACTIONS AND FACILITATION

The word 'Emotion' is occurred from the Latin word 'emovere', which means to stir up. It indicates a condition of being stirred up, aroused or moved in some way. Emotions is a state of feelings, impulses which is consisted of physiological and psychological reactions.

Emotion is the thinking of proper image to get a particular object or fulfill one's needs and be moved by observing it and The essence of emotion is motivation because emotional incentives create a tension in the entire activity of body by which a person's conduct is directed towards a particular object. This is a need of a person to express the proper emotional condition and we can know that emotional state by observation Emotions are important in human life. If an individual's emotions do not develop in a balanced manner, the individual's entire personality becomes disorganised, and many distortions manifest themselves. Emotions arouses in a person both constructive and destructive feelings. An emotion is temporary. An individual's emotions change too rapidly even for introspective awareness of the feeling. Sometimes the emotion lasts for a longer period, hours or even for a day or more. Then it is a mood. When a person is in an angry mood, even trivial things tend to prolong the emotion. A person in a happy mood may not get angry over certain irritations which would provoke a violent display of anger he had seen in an irritable mood.

A sentiment lasts almost permanently and influences an individual's reaction to an object or thing to such an extent that one can often predict his behaviour in any situation if one is aware of his sentiments. Emotion is an excitation approach. Emotions are disturbed states which originate in psychological situations and which are revealed by marked bodily changes in the glands and smooth muscles.

Aristotle asserted that human beings are rational. Generally it is believed that everything we do is done the rational way. This blinds us to the immense importance and significant of feelings and emotions in our life. The essential elements of our culture, are, drama, poetry, music, etc. have their foundations in the effective experiences of men. It is emotions and feelings that make us both human and humane.

Feelings and emotions permeate our social and personal life. The very rationality of behaviour depends to some extent upon the overall balance of our emotions. Affects and feelings, when they are mild and within certain proportions, enrich our lives with all their undertones and overtones. On the other hand violent, disproportionate feelings van plunge the individual and the society into untold suffering for which effective remedies are hard to find.

Emotion and motivation are closely connected. Both the words are derived from the same Latin root, *movere* meaning 'to stir', 'to agitate', 'to excite', ' to move'. Emotions are often defined as stirred up or disturbed states of the individual, which often move him into action. W. McDougall (1923) had described emotions as springs of energy. Instead of being merely passive phases of the mind as traditional psychology defines them, feelings and emotions are dynmic, reactive states of the organism. Emotions 'move' a man both in the sense that emotional situations move him to pity and sympathy and in the sense that they stir him up and impel him to act. Our everyday behaviour, both at home and in the society, is to a very large extent conditioned by our feelings.

It has been observed that emotional stability helps an individual in social adjustment. Education aims at bringing emotional balance in every child. Mc Dougall's classification of instincts and their corresponding emotions need to be discussed in brief.

The instinct of escape is aroused by the emotion of fear. Hearing the sound of bombardment the crowd not only take to flight but they make noise vigorously to warn other people to move away.

The instinct of combat is aroused by the emotion of anger. All children from birth scream and struggle is their bodily movements are seriously restricted.

The parental instinct is aroused by the emotion of tender. The mother is emotionally attached to her children. The appeal instinct is aroused by the emotion of disgust. We have observed behaviour, apparently indicating disgust, in the first year, on occasions when children encounter new foods. Their dislike appears to be due not only to the taste but also to the feel of the food their tongues, shown for example by the objection to lumpy porridge and to skin on milk.

The mating instinct is aroused by the emotion of lust. This emotion is aroused by the presence of a suitable member of the opposite sex and by the perception of the sexual organs of the opposite sex.

The instinct of curiosity is aroused by the emotion of wonder. It has been observed that the children first show curiosity in the presence of something unfamiliar but not totally strange. An example will make it clear. When the familiar object is missed from its accustomed place. Later, in the second year, they show curiosity in the presence of hidden things, as shown for example by their delight in opening drawers and cupboards, and by their desire to know the contents of parcels.

The instinct of submission is aroused by the emotion of negative self-feeling. The term 'submission' is reserved for behaviour that is the opposite of 'assertion'. It must not be confused with fear. Submissive behaviour is shown as early as the first year in the form of shyness-a desire to retire, to attract as little attention as possible. This seems to be an example of a child feeling his unimportance. The instinct of self-assertion is aroused by the emotion of positive self-feeling. A great deal of children's behaviour can be described as assertive. When they are angry or curious, they are in a sense asserting themselves. The same is true when they are making things or playing games. 'The assertive behaviour begins as early as the fifth month in the form of playful disobedience. The child is experimenting to see how far he can go in asserting his independence. The child also gives orders to other persons, he threatens them, he corrects them, he points out their shortcomings; he is asserting his superiority.

The instinct of repulsion is aroused by the emotion of disgust. We have observed behaviour, apparently indicating disgust, in the first year, on occasions when children encounter new foods. When very young children experience something obnoxious, they attend to it closely, their expression indicates nausea, and they spit out and thrust away the offending substance. We infer that they are experiencing the emotion which we call " disgust".

Characteristics of Emotion

Emotion is a hard term to define. When we speak of emotions, we usually refer to (a) subjective felling, (b) bodily states, (c) expressions of the emotion by language, facial expressions, and gestures, and (d) the motivational drive of certain emotions.

The expression and perception of emotion is quite important in our responses to other people. The tone and other characteristics of the voice are a channel for the expression of our own emotions and the perception of emotion in others. *Facial expressions* are perhaps the most important non-verbal way in which emotions are expressed. One early set of observation indicated that three dimensions of facial emotional expression can be judged with accuracy. These are the pleasantness - unpleasantness, attention - rejection, and sleep - tension dimensions. More recent studies, have shown that the facial expression of a number of primary emotions can be judged rather accurately by people from diverse cultures. This gives strength to the view that the expression of the primary emotions may be innate.

The relatively long-lasting emotional tone, or background, that colours our outlook on the world is called *mood.* Little is known about the specific situations giving rise to moods. In contrast to moods, a good deal is known about the specific situations that generate relatively short-lived and *episodic emotional feelings*, such as pleasure, fear and anxiety, anger and hostility, and depression and grief.

Pleasure is a reaction to the satisfaction of a motive or the attainment of a goal.

In general, *fear* is aroused by situations that are perceived as physically threatening damaging to one's sense of well-being, or potentially frustrating. Specific fears are learned through conditioning, through modeling, and by symbolic manes. Intense fears are known as *phobias.* In contrast to fears of specific things, *anxiety* is a vague fear experienced without our knowing exactly what the source is. Anxiety may be due to an unconscious memory of a fear situation caused by repression, or through stimulus generalization.

Anger and hostility are frequently reactions to motive frustration, injury, insults, and threats. The sources of anger changes with development, as do the ways anger is expressed. Anger conditioned to a particular situation or person can spread, or generalize, to similar situations or people; such generalized anger is called hostility.

Depression is often the result of the frustration of many motives. Depressed people lose the joy of living, feel inadequate and worthless, give up many activities, and may withdraw from other people. *Grief,* in contrast to depression, is an emotional reaction to a specific loss: it can be beneficial in helping a person come to terms with the loss.

Physiological Basis of Emotion

A number of bodily reactions accompany emotional states. Both the autonomic and somatic parts of the peripheral nervous system are active in strong emotions. The autonomic nervous system is the part of the peripheral nervous system, which regulates activitiy in the smooth muscles, heart muscle, and certain glands. The two divisions of the *autonomic nervous system* – the sympathetic system and the parasympathetic system – are active in many emotional states.

For a number of emotiøns, different patterns of bodily activity can be detected. Fear and anger are characterized by the emergency, or 'fight-or-fight', response of the sympathetic nervous system. But more fine-grained analysis of the bodily changes in fear and anger has shown that each has a somewhat different pattern of activity. The relaxation response, which accompanies clam, meditative states is a pattern of bodily activity that, so far as the sympathetic system is concerned, is almost the opposite of the *emergency response.*

The patterns of bodily activity in a number of emotions are controlled by the number of emotions are controlled by the *limbic system* and the *hypothalamus* of the brain.

The degree of arousal is an important part of many emotional states. One *indicator of arousal* is the electroencephalogram (*EEG*), or record of 'brain waves'. High frequency, low voltage activity of the cerebral cortex indicates arousal; low frequency, high-voltage activity indicates sleep or a low state of arousal. Other indicators of arousal include the galvanic skin response (*GSP*), *papillary dilation*, and th *orienting reaction*.

For complex tasks, the relationship between arousal and performance is an inverted U-shaped curve. Accuracy of performance is low at low levels of arousal, optimum at medium arousal levels, and low again at very high levels of arousal.

The *reticular formation of the brain stem* and the fibers ascending from it to higher portions of the brain make up what is called the *ascending reticular activating system (ARAS)*. It is this system which is responsible for controlling levels of brain arousal, and hence, to a large degree, behavioural arousal.

If strong emotional reactions persist, the bodily changes that accompany the emotions can contribute to various *bodily disorders*, such as high blood pressure and ulcers. These bodily disorders, or disease, are termed *psychosomatic reactions*.

The bodily changes in emotion can sometimes be used to detect whether people are lying. The fact that telling a lie upsets most people, thus producing bodily reactions, is the rationale behind the use of *"the detector"*, or *polygraphs*.

Influence of Emotion on Performance

Emotions are a part and parcel of the human behaviour. Emotions play an important role keeping the individual healthy. Emotions are a great 'reservoir' of energy. It has been seen that unless an athlete is emotionally aroused, puffed up, and charged with psychic energy, he/she will not show excellence in games and sports. The coach has the responsibility of 'arousing' players to dead-pitch so that they win and not get defeated. Some athletes are found to be physically tensed as regards lack of their physical skill. Such emotional states like fear etc., the coach should feel sympathetic towards the learning athlete and remove the inadequacies. The performing athletes have to be taught as to how to 'play cool', to face competition with the nerves of steel and to remain normal under tough situations.

It is interesting to note the complex emotions that result from the blending of various simpler emotions; e.g. disgust and anger produce scorn, if assertion is added the scorn changes to contempt. Disgust and fear produce loathing. Wonder and submission give rise to admiration. Conversely, complex emotions, such as envy, jealousy, gratitude may be received into their components.

An athlete's emotional behaviour profoundly influences his/her social development. An obstinate, irritable and easily angered athlete does not easily adapt himself/ herself to a group. In fact, it is no exaggeration to say that emotional and social development are complement to each other.

The athlete's social development is influenced by his/her games and his/her companions. Games develop the feelings of cooperation, sympathy, competitiveness, leadership, etc. The social and economic status of the parents has also an impact upon the athlete's social development. Children/athletes of poor parents play games such as hide and seek or 'kabbadi' because these do not involve any expense, whereas children/athletes from rich homes play games requiring considerable expenditure. This is one of the most important factors influencing social development. Thus we see that the influence of emotion of the social development and performance play an important role. In this a significant role is played by the family group, primary and secondary groups, society, nation, etc. Many new facts have been discovered in the field of sport psychology as a result of studies conducted on socio-emotional development.

Conflicts : A conflict is a situation in which an individual is required to act in two incompatible ways to achieve two or more mutually exclusive goals. In a conflict you are threatened with frustration regardless of the course you choose to follow. It is different from frustration or anxiety, in that it involves decision making. The frustration or anxiety is relates and restricted to the decision making period. In a conflict, the contradictory alternatives are approximately of equal attractiveness.

The selection of one means the denial of the other. The individual therefore hesitates and vacillates during which period he/she undergoes considerable stress and strain. Conflicts are of three types-(a) approach-approach conflict, (b) approach-avoidance conflict, and (c) avoidance-avoidance conflict.

(a) In approach-approach conflict, there is a competition between two alternatives, both of which are equally attractive and desired. The individual wants both but cannot have both because of some limitation in time, energy or other circumstances. A youth who is faced with two offers for a job after a long waiting both the offers being equally attractive, a student who is faced with a choice of taking part in a drama or a debate held at the same time, both of which hold promise of success, are examples of an approach-approach conflict.

(b) In the approach – avoidance conflict, there is a strong tendency to both approach and avoid the same goal or need. In this case the object or the need is such that it is appealing on the one hand, but at the same time involves some undesirable result. The individual must take the bitter along with the sweet. If he accepts it, he accepts a little satisfaction and a little of pain.

(c) Avoidance-avoidance conflict involves a choice between two alternatives both of which are unattractive. The saying caught between the devil and the deep blue sea is apt here. To accept one of them is a must, whether the individual likes it or not; for instance, an idealistic principal may be confronted with the choice of doing something against his/her conscience facing the threat off dismissal. Another example could be of a highway traveller accosted by bandits. He/she is threatened with the loss of his/her belongings or loss of his/her life.

Frustration : It is the feeling of being blocked or thwarted in satisfying a need that the individual perceives as significant. In other words it is a blocking of an individual's strivings. Frustration is a conflict situation in which the individual must complete one task or turn to another. In the coaching camp athlete faces frustrating situations when either he/she wants to or feels that he/she must engage in activity/game which competes with his/her achievement of the instructional objective. The coaches face frustration when they must engage in activities which compete with the attention they could give to their coaching objectives.

Research on frustration interestingly enough indicates that the frustrating (or conflict) situation may sometimes increase individual's efforts to compete the task. When this occurs, the frustration becomes a challenge the individual increases his/her determination to finish what he/she has started.

Whether frustration assists athletes in achieving instructional objectives depends on their entering behaviour. Frustration in almost every case will invigorate athlete behaviour but it will not necessarily cause the athlete to invest his/her effort in practising tasks. It will elicit those responses in the athlete which have the greatest strength. Since frustration is a part of the human condition and none of us escapes it, the coach must find ways of channeling the effects of frustration toward achievement rather than toward either non-achievement or the achievement of alternative or irrelevant objectives.

Anxiety : Anxiety describes the individual's level of emotionality. We can sometimes distinguish among our friends who are mainly tense and worried (highly anxious) and those who are cool (hardly anxious). We know that anxiety and arousal are relates because at the higher levels of arousal we considerably have more emotionality than at the lower levels. Since anxiety is an inferred emotional state of the organism and cannot be directly observed, investigations of anxiety rely heavily on having the individual report his/her own emotional states under various stress conditions. Sports psychologists have studies test anxiety, or emotional states the athlete experiences under the stressful conditions of taking a test. These investigators developed a questionnaire, the Test Anxiety Scale (TAS), to measure anxiety in test situations. In responding to the questionnaire the athlete reports such subjective reactions as worry before and during examinations or athletic competitions, heart palpitations, uneasiness, and sweating. These findings emerged from the various studies : (1) Low-anxiety athletes perform better when challenged by the task and by the prospect that their performance will be assessed; (2) High-anxiety athletes perform worse under these same conditions. They perform better when they are not threatened with evaluation and faced with a challenge.

It has been observed that anxiety is a physiological response to a real imagined threat. It is a complex emotional state characterised by a general fear. Feelings of rejection and insecurity are usually a part of anxiety. A certain amount of anxiety is needed for peak performance.

SOCIAL INTERACTIONS AND FACILITATION

Social facilitation is consisted of the effects of co-action which is generally considered as beneficial. Various Psychologists and Behaviourists used this term when they theorized that social facilitation should be further subdivided into audience effects and coaction effects. They felt that a real distinction exists between the effects that the mere presence of spectators have on performance and the effects due to other individuals performing the same task. Competitive behaviour will be higher or stronger in a coaction situation than in a mere audience situation. The circumstance that trigger are believed to be present in the environments of the athlete which interact with his internal make up influencing the individual's arousal level. It has been observed that in majority of the sports areas, social evaluation of an athlete's performance acts as the driving force in this regard. The above discussion reveals that social facilitation refers to any increment of individual behaviour resulting from the presence of another individual. Neumann, for example, noticed that each time he entered the laboratory his students would markedly improve their performance on a finger ergograph (a device for measuring and recording muscular work competed) as compared to when they worked alone. Social facilitation explains only one part of the influence of the presence of other individuals. A number of early investigations also found social impairment effect. For example, the presence of several spectators distributed and interfered with the learning of nonsense syllables and a finger maze.

Individual Variables : Under this category one variable receiving attention has been the personality disposition of anxiety. A study conducted by Cox found that high anxious children decreased their rate of performance and low anxious children increased their rate of performance when in the presence of others. Cox suggested that the presence of another person was interpreted as transforming the situation into as evaluative one for high anxious children, causing their anxiety to interfere with performance; for low anxious children an observer was assumed to provide an incentive for more efficient performance. Observer presence was detrimental for high and moderately anxious females. Martens found no evidence that a person's anxiety level mediated the audience effect on a motor task.

Audience Effect : Martens conducted a study and used a coincident timing motor task and college males as subjects and audience members. The results clearly showed that when learning a complex motor skill (analogous to hitting a baseball), the presence of ten passive spectators impaired the subject's initial performance as compared to individuals learning alone. On the other hand, after the skill had been reasonably well learned, individuals performing in the presence of an audience were superior to those performing alone. Significant increases in palmar sweating occurred among the individuals learning and performing in the presence of an audience as compared to those learning and performing alone, indicating that the presence of others was arousing.

In practical terms, apart from his inherent personality makeup, specific variables operate in a social situation to influence a person's behaviour and performance. Competition situations consists not only of cognitive desire to win (rivalry) but also of a general effect called social facilitation. This social facilitation (or raising of arousal level) is a function of both coaction and audience effects.

Presence of others : Zajonc popularised the term 'social' facilitation' as a general term referring to both positive and negative effects as a consequence of the presence of others, although its initial use in this context should be credited to Allport. Zajonc was able to provide some evidence that the presence of others in creases arousal. He concluded that the mere presence of others impair learning because the aroused individual more frequently emits the incorrect response, but facilitates the performance of a well-learned task.

Coaction Effect : The researches conducted by psychologists indicated that the subjects tolerated more shock in a group situation than alone. The ability to ensure more pain to sustain a physical effort is an important part of achieving excellence in sport. Seidman's study suggests that the perceived sharing of stress contributes to stress tolerance. Carment found that coactors increased the rate of responding on a simple motor task. The study conducted by Ader and Tatum, using a motor task, provides any evidence that learning is impaired when in the presence of coactors. Thus we need to determine if the presence of coactors impairs the learning of complex motor skills.

Factors Mediating the Social Facilitation

A variety of variables may determine precisely how the presence of evaluating others affects any particular individual. These factors fall into the categories of individual situation and task variables. These factors are discussed below.

Situational Variables : The increasing thing in section is the relationship between the "characteristics of the audience members or coactors and the performer. For example, does a supportive audience influence a performer differently than a non-supportive audience? In a study Kozar failed to find out any differences between these two types of audience when performing a gross motor balancing task. Other variables that must be taken into account are age and sex differences between the performer and the audiences. The size of the audience needs further attention. The formality of the situation may heighten or lessen arousal manifested by the presence of others. Finally, it would be helpful to know how previous experience of others influences performance.

Task Variables : The presence of others will facilitate performance on a simple task and impair performance on a complex, unlearned task. Another factor that we need to consider is the relationship between the audience or the coach-elicited arousal and performance on fine and gross motor skills.

Methods of Investigation in Sport Psychology

The testing tools of sport psychology have come from psychology. Efforts have been made to prepare the sports psychological tools. The sporting situations are difficult than other normal life situations. Hence, gradually such tests have been prepared. Still we are using two methods namely observation and introspection. It is better to discuss all the methods taken from psychology. In studying the problems of sport psychology, attention is paid to the facts that they must be free from error and capable of giving reliable and valid results.

The Armchair Method : This method depends upon a minimum of observation and a maximum of speculation. Psychology as a science cannot accept this method which is based on mere speculation.

Method of Systematic Observation : Suppose a teacher wants to know how children at various age levels express anger. One way of doing this is to seek the cooperation of mothers by giving them necessary training in systematic observation.

A number of problems can be studied by following this method; the following illustrations seem to be useful.

1. When does a child utter the first word?

Which word is it?

2. When does the child begin to play with other children?

Answers to these questions are obtained by giving training to parents and to coaches in systematic observation. This method is useful for studying the behaviour of animals also. Observation is done under natural conditions. It cannot be done by creating artificial conditions.

The Experimental Method : The most dependable way of obtaining facts is the experimental method. It is used to investigate behaviour which can be brought into the laboratory and studied under controlled conditions. Let us consider the following problems :

1. What is the effect of incentives on the achievements of the athletes?

2. How do the blind perceive obstacle?

3. Does smoking lower mental efficiency?

4. How does job satisfaction affect the performance of coaches?

In a typical experiment there are at least two group namely, an experimental group and a control group. The two groups should match one another in respect of age, ability, motivation etc. For example, in case of problem 1 above, there will be three matched groups. They will be taught a particular skill by the same coach and by the same method. The only factor that will be different in the three groups is the incentive. This is known as independent variable. The experimenter (in this case the coach) may use praise as incentive in one group, blame as incentive in the second group and no incentive in the third group. The achievements of the athletes which are likely to depend on the incentive is known as dependent variable.

In the experimental method, sporting experiments are conducted in a controlled environment. In such an environment we can find psychological solutions to the various problems of sport. In employing this method, some facts or problem is taken up for experiment. All facts or materials likely to conduct the experiment are kept away from the laboratory. In this way, the entire environment is rigidly controlled. The activity which obstruct the experiment is related to mental activity. The observer acquaints the subject with the former activity and then invite the subject to enter into the pure mental activity. During the test, efforts are made to understand the subject's interest, memory, attention, rational thought, mind and the various processes of behaviour, their nature and complexities. In the experimental method, it is possible to measure quality as well as quantity. It is a method by which the effect of a single designated factor is studied by applying it to one individual or group and not to another or by varying the factor under controlled condition.

Method of Introspection : Introspection means looking into oneself; that is self-observation of one's own mental states. It is something we are doing almost constantly in our daily life. Let us consider the following remarks :

1. Sanjay tells the doctor, "I an feeling well now."
2. Nanda tells her friend, "I was terribly afraid

yesterday night."

3. Raju says, "I feel confident to win the match."

All these remarks indicate that the persons concerned are telling something about their own mental state. They are observing and then reporting their own experience.

Three clear stages can be distinguished in introspection :

(1) During the observation of an external object, the person begins to ponder over his own mental states. For example, while listening to the music, which is to him pleasant or unpleasant, he starts thinking about his own mental state.

(2) The person begins to question the working of his own mind. He thinks and analyses : why has he said such and such a thing? Why has he talked in a particular manner and so on.

(3) He tries to frame the laws and conditions of mental processes. He thinks in terms of the improvement of his reasoning or the control of his emotional stages. This stage is that of the scientific methods for the advancement of our systematic knowledge.

The Case History Method : In this method, the total history of the individual/subject is traced. The individual's previous history is gathered from him and his associates and on the basis of this, certain conclusions are drawn. This method is mostly used with criminals or with those who show some sort of anti-social behaviour. Its use involves great caution as the people are likely to distort the data. This method is scientific, reliable and valid. In this method, an account of subject's life cycle is obtained. After collecting the individual's history, the psychologist studies the factors influencing him as well as the shortcomings in his personality. The observer wins the confidence of the children and then tries of find out facts pertaining to their past. He tries to find out facts relating to the child's neighbourhood, his circle of friends, parents, guardians and condition of the family. After discovering the totality of influence, he then forms an opinion about the child and provides remedies for the child's problems.

Regarding case history, the researchers opine that it is a suitable method for diagnostic studies. It provides complete information about individual's behaviour, his environment, insight to the nature of the case and for diagnosing the cases of suffering personal social, emotional and educational abnormalities. This method is used for a single person and its aim is diagnosis and treatment. Little Judy is throwing temper tantrums, not eating her meals, crying all night and refusing to go to school. Deepak, who is otherwise a fine young boy, is caught stealing some articles from a shop. Rahul is worried as he and his wife cannot get along. Sushila wants to know the career that is most suitable for her. All these cases are having problems. They approach a counselor for help.

Clinical Method : This method is generally used in the study of specific learning, behavioural disorders and complexes. In it, various clinical methods and techniques are employed in conformity with the problem under examination. Their objective is to recognise or discover the particular needs of the subject, the causes which give rise to the complex and the assistance that should be given to the subject.

Skinner has opined that the clinical method is most commonly employed in the study of complex problems related to learning, personality and behaviour. For this purpose of using this method is to find out what the particular needs of the individual are to identify and understand the causes of the abnormalities and complexities generated in him and to discover how the patient can be helped.

Comparative Method : Comparative methods is a system of study which aims through the examination of existing facts or those which have existed in the past to assemble a definite body of material from which the investigator, by selection, comparison and elimination may discover the idea type and progressive force of the events. It is a system of study which involves the collection, classification, analysis, and description of theory and practice of the facts and principles. This method is brought to use when one individual's behaviour is to be compared with the behaviour of other individuals. Comparison helps us to understand many facts and similarities and consequently, the truth becomes apparent. This method is useful in comparing two types of behaviour and increase the knowledge by investigating the similarities and differences among the subject.

Psychoanalytic Method : Sigmund Freud is the father of this method. In this method an effort is made to understand the innermost thoughts of the individual. According to Freud, 7/8 th of the mind remains in an unconscious state while only 1/8 th part of it remains in a conscious condition. This method aims at studying the unconscious part of the mind. This method is credited with many of the major achievements is psychology. It takes the help of dream analysis, word association, free association and hypnotism. Jung and Adler developed this method in three stages namely, (i) establishing rapport with the subject; (ii) transference and (iii) breaking the transference. Elucidating the limitations of the psychoanalytic method, Woodworth has stated that this method requires a lot of time. This method should not be used till the patient himself/herself is willing to undergo treatment till the very end. If the treatment is suspended before completion, the patient's condition becomes considerably worse. Even psychoanalysts do not consider it a method which can provide cure but it has nevertheless been observed that the behaviour of patients treated by this method becomes normal.

Statistical Method : It becomes necessary to take the help of statistical method to make the study of sport psychology reliable and valid. In this method, the data obtained is analysed and the reliability and validity of conclusions is determined. Statistic is a value or measure that describes a particular series of quantitative observations. The universe from which the observations were drawn to estimate the corresponding value in the universe. It is a summary of value based on sample. It is a term sometimes used to designate any statistical item, observation, score measure or similar datum.

Genetic or Developmental Method Through this method, a systematic study is made of the development of the individual or group. Through this study, the various stages of development, the features of development and its effect upon the human personality are studied. Problems relating to environment and heredity are studied by this method.

In this method, the technique of observation is employed over the period of growth. In this method we need to study individuals day by day, month by month and year by year, noting the changes that occur in them. Many types of tests are used to study the growth and decline of various abilities such as numerical, verbal and spatial ability, reaction time, memory, reasoning etc. The method of systematic observation is also used in the genetic method. At a nursery school you may find a psychologist jotting down notes about Swati or Nagina day after day. A one-way vision screen sometimes separates the child from the observer so that the child may not become aware of the psychologist's presence.

SPORT SPECIFIC PSYCHOLOGICAL TESTS

Psychological testing is the most outstanding contribution that comes solely from the efforts and constant research by the sport psychologists. Its value is immense in the sense that tests are the most sophisticated tools of measuring human characteristics and individual differences that form the basis of sport psychology.

Testing is a highly complex sub-field of modem psychology in general and sport psychology in particular. In using any sound measuring instrument the first question that has to be answered is : Does the device (test) really measure what it is intended to measure? If the test measures and serves the purpose of the intention of the test, it is considered valid. A necessary condition for validity is that the test must give scores that are consistent. The same subject must score in about the same way if he/she were tested again by the same instrument under similar conditions. An instrument that gives consistent scores is called reliable. Inches, foot-rules are, in this sense, very highly reliable measures, because they will give precisely the same measure of the same aspects of the same things even if they are repeated hundred times. Thus, the reliability of a test may be defined a test's ability to give consistent results when repeated.

In order to be reliable, the measuring instrument must be objective. It should be kept in mind that if two or more persons use it on the same person, the result of, test must also tally again, if the subject's test score is to impart any meaningful information, it must be interpretable in relation to scores made by the other individuals in a defined group. This is accomplished by 'standardization'. Thus, the most distinguishing marks of a sound scientific psychological test are that it must have a very high validity, reliability, objectivity and its results must be clearly interpretable.

Importance of Psychological Tests : The importance of sports related psychological tests is fairly evident in the selection and placement of athletes. Defining psychological tests, Marshell writes, " A psychological test, then, is a pattern of stimuli selected and organised to elicit responses which will reveal certain psychological characteristics in the person who takes them." In this way, psychological tests are organization of some stimuli. These stimuli are selected with some special objectives. The object is to discover some psychological characteristics of the sportsperson being tested. Stimuli present in psychological tests are called test items. These can be of several kinds namely, pictures, figures, wooden blocks, word group etc. The way in which these stimuli uncover psychological characteristics will be evident from a description of each of them.

Characteristics of a Good Test

Good tests of a psychological nature bear the following features :

1. Reliability : The higher the degree of reliability that results of test have the greater is considered the reliability of the test itself. The reliability of its conclusion means that if the same test is performed upon the same athlete more than once, results obtained in each athlete list should be the same or commensurate with the first one, if it is modified by changed circumstances.

2. Validity : The validity of a test lies in the fact that it succeeds in testing the particular quality which it is designed to test. For example, intelligence tests can be valid only if they test correctly the intelligence level of the individual subkect/athlete.

3. Objectivity : A test can achieve a high level of objectivity only if the answer to each of its various questions is distinct and precise and if there is no difference of opinion regarding it among the various examiners. Its efficacy lies in the fact that the result is in no way influenced by the examiner's individual thoughts so that the results are obtained no matter what the number of examiners who examine the same subject.

4. Comprehensiveness : Another feature of a good

test is its comprehensive nature. This feature is inclusive of all features allied to the one in question. It will provide information of qualities concerned with the one quality actually being tested. In this manner, in a comprehensive test, questions cover the quality in question as well as allied qualities so that no aspect of the measured sports ability escapes attention.

5. Utility : This feature concerns the extent to which a particular test can be used with facility. For this, it is essential that the questions asked should be simple, easy and less time consuming. The expenses are negligible.

Of the above mentioned characteristics, the most important are reliability and validity.

7

PERSONALITY

Personality is defined as 'distinctive patterns of behaviour (including thoughts and emotions) that characterize each individual's adaptation to the situations of his or her life.

Personality is shaped by inborn potential as modified by experiences common to the culture and sub-cultural group (such as sex roles) and by the unique experiences that affect the person as an individual.

Sport psychology studies sporting behaviour. Behaviour is he manifestation of personality. The manner in which a person's behaviour shows the kind of personality he possesses. " Personality is the whole man, his inherited aptitudes and capacities, all his past learning, the integration and synthesis of these factors into characteristic behaviour patterns, his ideals, values and expectations," Adequate and proper development of an athlete's personality reinforces his/her capacities and potentialities.

The word ' personality' has been derived from the Latin word 'Personae' which means " to sound through." This term was used to describe the voice of an actor speaking through mask. This term slowly began to be applied to the actors themselves. By personality, it is now generally meant that it is the organising and integration of a large number of human traits. The concepts of personality differ widely among different people. Personality may mean different things to different people. To some, it means a general sum of traits of the person. To others, it refers to unitary mode of response of life situations. Thus there prevails a great deal of controversy over the meaning of personality.

Psychologists have defined personality in different ways. Valentine's view is the sum total of innate and acquired' dispositions whereas Boring's view deals with 'an' individual's adjustment to his environment. Morton views personality as the 'biological innate dispositions, impulses, tendencies, aptitudes and the instincts of the individual'. Munn views personality as the 'integration of an individual's structures, modes of behaviour, interests, attitudes capacities, abilities and aptitudes'. Rexroad explains personality as "the balance between socially approved traits". Dashiell views personality as his system reactions'.

As one observes sport and physical activity, whether as an athlete, coach, trainer, team physician, or spectator, it becomes quite obvious that participants from various athletic sub-groups differ in a number of respects. Indeed, there is a wealth of research literature attesting to the differences in physiologic and somatotype differences of athletes from various sub-groups. While observers of sport seem to be equally confident that such differences have psychologic correlates, there is actually little agreement as to the psychologic ways in which these same athletic sub-groups differ from each other or from nonathletes. The purpose of this section shall be to (1) review the existing literature concerning the athlete's personality, (2) attempt to explain the lack of agreement which seems to exist on this topic, and (3) offer suggestions for research strategies to be employed in the future.

Interpretation of the literature dealing with athlete's personalities might be approached from a number of perspectives. However, the author feels that it is efficacious first to examine this literature from the standpoint of the independent and dependent variables which have been used historically.

Independent Variables. First of all, numerous independent variables have been employed, and furthermore, various operational definitions have been used for the same variable by different investigators,

(2) athletes and nonathletes,

(3) team and individual sport athletes,

(4) combative and non-combative,

(5) the same athletic sub-groups from different types of educational institutions, and

(6) athletes of differing ability levels. Also investigators such as Goals have compared athletes from team and individual sports and then reassigned the athletes into combative and non-combative sport groups for comparative purposes.

[A] general problem has been that investigators have failed to employ universally acceptable definitions of group affiliation. The general practice has simply consisted of testing intact athletic sub-groups, with no attempt to differentiate between ability levels. The athlete who "rides the bench" throughout his athletic career might differ in a number of ways from the "starter".

In the other words, while athletes from one sub-group may not differ from those in another, it is quite possible such groups would differ both within and between sports. In view of the fact that investigators have failed to employ adequate operational definitions of the independent variables used in this area of inquiry, it is not surprising that general confusion seems to exist. It is clear, of course, that subsequent investigations must include rigorous operational definitions of the independent variable.

Dependent Variables. The assessment of personality can be pursued with either direct or indirect methods. The majority of investigators in the field of sport psychology have elected to employ direct methods which consist of self-reports such as the Edwards Personal Preferences Schedule, Eysenck Personality Inventory, Minnesota Multiphasic Personality Inventory, and the Sixteen Personality Factor Questionnaire.

Several investigators have employed indirect procedures which are unstructural. Examples of these projective techniques are the Rorschach Test, Figure-Drawing Test, Thematic Apperception Test, House-Tree-Person Test, Sentence Completion Test, and the Rosenzweig Picture-Frustration Study. The use of direct methods of personality assessment as compared to indirect techniques as well as the efficacy of employing a battery of both direct and indirect procedures in sport psychology have been discussed previously.

In view of the nosologic problems associated with projective methods, it would seem that sport psychologists have chosen the correct approach, that is, direct techniques facilitate within and between laboratory replication. Unfortunately, investigators have employed various personality inventories such as the EPI, EPPS, MMPI, and 16 PF, and then attempted to make direct comparisons with one another's work. The traits measured by these inventories are essentially psychologic constructs that have been derived from factor analytic work.

The output of such analyses is dependent in large part upon the input. Also, were factor analysts to begin with same input, subsequent rotation and labelling might differ markedly. On the hand, independently constructed inventories such as Cattell's 16-PF, Guilford's personality inventory, and the Eysencks' EPI measure the dimensions of extroversion-introversion and neuroticism-stability "with almost complete agreement" .

At the same time, while the MMPI contains a measure of extroversion-introversion it was not designed for the measurement of E in "normal groups and it is not well adapted for this purpose". Also, fractionation of the MMPI is hazardous and the specific scales should be examined in light of scores on other dimensions of the MMPI. At any rate , while it is difficult to make direct comparisons of investigations which have used different inventories, this is certainly possible, if the investigator has an understanding of the various instruments and the recommended procedures for the their administration and interpretation. Rather than simply review investigations which have dealt with the psychologic characteristics of the athlete, the author has elected first to comment on the methodologic problems which have historically plagued this area of inquiry.

One of the more obvious problems has been the general inadequacy of operational definitions of the independent variables used in these investigations. Also, while more acceptable operational definitions have been employed for the dependent variables, the variety of instruments employed does not make for ready comparisons. These points should be kept in mind when attempting to interpret the literature in this area.

PERSONALITY CHARACTERISTICS OF ATHLETIC GROUPS

The present review deals with investigations which have been directed toward an understanding of the personality characteristics of a various athletic sub-groups. Studies have conducted on participants in baseball, basketball, cross-country, football, golf, karate, marathon running, parachuting, swimming, scuba diving, tennis, track, weightlifting, and wrestling.

Also, psychoanalytic observations of sport have been made by Adatoo, Beissetr, Deutsch, Harlow, and Whitman. In addition to evaluating athletes from different sub-groups, investigators have also compared athletes and nonathletes, athletes from different types of educational institutions, and athletes of differing skill levels. Also, numerous psychometric tools have been employed in these studies. Therefore, any categorization of this research for discussion purposes must obviously be arbitrary.

In most instances there has not been sufficient research on specific sports to justify a review. It does seem, however, that several common research strategies have been adopted by investigators, and these have been used in developing a topical outline for this review.

Sampling

Most investigators have tested readily available, intact groups of athletes. The assumption has been that participants in a given sport in one educational institution or geographical locale are similar to those athletes in the same sports in other schools and areas.

There is evidence which suggests that athletes from different sports have different somatotypes. Also, a gymnast or diver at one school would not be expected to be a football tackle or heavyweight wrestler at another institution.

Therefore, if Sheldon's theory of personality were adopted, this sampling technique might not be as dangerous as it seems. At any rate, there is evidence which suggests that sampling is a very important consideration in attempting to explain the athlete's personality. This point is well illustrated in an investigation by Morgan who administrated the Eysenck Personality Inventory to English-speaking wrestlers participating in the 1966 world championships.

The group consisted of the Canadian, South African, and United States teams. When viewed collectively, this sample was stable, but somewhat more extroverted than normal. However, when the teams were evaluated independently the United States team was found to score significantly lower on the neuroticism-stability dimension than the South African team. In short, a somewhat typical profile was observed at the collective level which actually clouded the rather unique characteristics of these two sub-groups.

Athletes and Nonathletes

Investigations which have been concerned specifically with comparisons of athletes and nonathletes will be reviewed in this section. Henry administered a questionnaire consisting of items from the Thurstone neurotic inventory and questions pertaining to ascendence-submission to student pilots, members of a college track team, physical education major students, and a group of students enrolled in weightlifting.

The tracks athletes and pilots were found to be quite similar and these groups were significantly less hypochondriacal and introverted than the weightlifters. Also, they were significantly more neurasthenic than the physical education majors. Slusher administered the MMPI to high school who had won letters in baseball, basketball, football, swimming, and wrestling, and to nonathletes from the same population.

The athletes and nonathletes differed on all of the MMPI scales except the M and K. These MMPI scales are known to fluctuate with various treatments. For example, scores on the depression scale decrease following the administration of antidepressants and numerous other therapies. At any rate, since the MMPI is not comprised of trait measures per se, it cannot be inferred that the athlete and nonathlete groups differed prior to the athletes electing to participate in sport. Schendel administered the California Psychological Inventory to ninth-grade, twelfth-grade, and college athletes and nonparticipants in athletics.

The ninth-grade athletes differed from the nonathletes on eight of the CPI scales, differences existed on 4 scales for the twelfth-grade sample, and nine differences were observed for the college Ss. This investigation supports the view that athletes and nonathletes differ in personality structure, and while cross-sectional studies have younger and older groups differed. The Gordon Personal Profile was administered to 22 Negro and 35 white athletes and 19 Negros and 35 white nonathletes by Hunt. He reported that Negro and white varsity athletes had similar personality profiles as did the Negro and white nonathletes.

Also, the athletes, regardless of ethnic background, tended to differ from the nonathletes. The Edwards

Personal Preference Schedule was administered to 950 males freshmen by Fletcher and Dowell. These subjects were further divided into groups who had participated in high school athletics and those who had not. The two groups were found to differ on the Dominance, Aggression, and Order scales of the EPPS. College wrestlers and experienced marathoners have been found to score significantly lower than the population average on anxiety.

Also, American world-class wrestlers were found to be more extroverted than the normal population and marathoners, who in turn were more introverted than the normal population and most other athletic subgroups. These investigations as well as the majority of the studies reviewed in this section suggest that athletes represent a published norms for the "normal" or "average" population if, in fact, athletes represent another population; that is, extensive mourning on athletes might be appropriate.

The Eysenck Personality Inventory was administrated to 46 team sport and 36 individual sport athletes as well as 40 nonathletes by Goals. The 82 athletes were also classified as representing either a combative or noncombative sport. Regardless of classification procedure the athletes were found to be significantly more extroverted than the nonathletes.

The sport groups did not differ on the E dimensions, however. Also, none of the groups differed on the neuroticism-stability dimension. One consistent finding throughout the sport psychology literature has been the observation that athletes tend to be normal on the neuroticism-stability dimension of personality. Indeed, American wrestlers were noted to be extremely stable even when tested 24 to 48 hours before competition in a world tournament.

It may be well be that stability is a prerequisite for high-level competition. A report by Yananda and Hirata suggests that this observation may reflect selective mortality. They reported that those students who dropped out and those who continued in their athletic clubs scored within the abnormal and normal ranges respectively on the Tokyo University Personality Inventory. Those students who continued in their sport clubs were less neurotic, less depressive, and more hypomanic than those who dropped out.

A different interpretation, however, might follow from the work of Carmen, Zerman, and Blaine, who compared the use of the Harvard Psychiatric Service by athletes and nonathletes. Athletes used the service less frequently, but those athletes who requested treatment tended to have more problems than did the nonathletes. Also, Pierce reported that athletes are less likely to make use of campus psychiatric facilities than are other students.

Of further interest was his observation that differences between athletes and nonathletes were "related to ability to take the role of patient than to actual degree of pathology, and the suggestion is made that ways of helping these legs verbal, less intellectually oriented students be developed and used". Little compared athletic and nonathletic neurotics. There has been an absence of neurotic markers in the life histories of the athletes, whereas neurotic markers were quite common in the nonathletic group.

Also, the athletic group was highly extroverted and sociable while the non-athletic group was characterized by introversion and a lack of sociability. Despite histories of good mental health, "the prognosis under treatment for the athletic group was, in general, less favourable". Neurotic breakdown were associated with a threat to physical well-being in 73 percent of the cases or the athletic group, while the percentage was only 11 for the nonathletic group.

The implication here for directors of athletic programs is that psychologic first aid is just as important as physical treatment for the injured athlete. Little's findings are essentially in agreement with those of Carmen, Zerman, and Blaine , and Pierce; that is, athletes display fewer neurotic symptoms than nonathletes, but prognosis for the athletes is not favourable once he is referred to a psychiatric service. On the basis of this review as well as the material contained in the previous section, it seems reasonable to conclude that athletes have consistently been found to differ from nonathletes on a number of personality traits.

The athletes tends to be stable and extroverted, with the exception of cross country runners and marathoners who are characterized by introversion. There is some evidence which suggests that even though they might be in need of such attention, athletes are less likely to seek psychiatric service than nonathletes. Once the athlete arrives at a psychiatric service, however, it appears that his prognosis is less favourable than that of the nonathlete.

Ability Levels

In addition to comparing athletic sub-groups several investigators have directed their attention toward an understanding of the relationship between personality and level of performance. For example, in one the earliest investigations Hohnson, Hutton, and Johnson administered the Rorschach and House-Tree-Person

Test to 12 athletes from contact and noncontact sports who were either of All-America or national champion calibre. These outstanding athletes were found to possess several distinguishing characteristics. These characteristics included "extreme aggressiveness, a freedom from great emotional inhibition, high and generalized anxiety, high level of intellectual; aspiration, and feelings of exceptional self-assurance". Another investigation of outstanding athletes was conducted by La Place who administered the MMPI to 49 major league players who served as his "successful" group and 64 minor league players who represented his "unsuccessful" group.

The major league players scored significantly lower than the minor league players on the Schizophrenia and Psychopathic Deviate Scales. Both groups, however, scored within one standard deviation of the mean for the general population, and therefore, it would appear that neither group differed appreciably from the normal.

This investigation revealed that the major league players were better adjusted than the minor league players. Whether these differences existed prior to each group's entrance into professional baseball and hence played a role in determining success, or whether these differences were a consequence of achievement levels cannot be inferred. A related investigation was conducted by Singer who administered the Edwards Personal Preference Schedule to 69 athletes at Ohio University. The athletes consisted of 26 varsity baseball players, 33 freshman baseball players, and 10 varsity tennis players. The ability level of each athlete was ranked by his coach.

Comparisons were made between the baseball players, tennis players, and the EPPS norms. Also, the comparisons were made between the highest, and lowest-ranked athlete in both sports and the EPPS norms. The baseball team scored significantly higher than the other two groups on the Abasement factor, significantly lower than the other two groups on the Intraception variable, lower than the tennis group on the Achievement variable, lower than the norm group scored significantly higher than the norm group on the Aggression factor.

No differences were noted between high- and low-rated baseball players, and the high- and low-rated tennis players only differed on one of the 15 measures. This investigation would be difficult to replicate since ability was not assessed with a universally recognized tool. Also, one would expect differences on numerous variables to be greater between major and minor league players than between high-and low-rated athletic groups from the same university team.

Similarly, high-and low-rated baseball and tennis groups from the Big Ten Conference might differ on a number of the EPPS scales, but such psychologists employ clear distinctions of ability in testing high- and low-performance groups. The MMPI was administrated to freshman and upper-class athletes and nonathletes by Booth. Comparisons were made between (1) athletes rated as poor or good competitors, (2) athletes who participated in team, individual, or team-individual sports, and (3) freshman and upper-class athletes and nonathletes.

The results revealed that the athletes and nonathletes and the athletes from the various sport groups differed significantly on several of the MMPI scales. Also, an analysis of the 550 MMPI questions revealed that 22 of the items discriminated between poor and good competitors. Kroll administered the 16 PF to 94 amateur and collegiate wrestlers consisting of 28 superior athletes who had been on the United States Olympic team or had been NCAA or NAIA champions or place winners, 33 college wrestlers who were rated as excellent by their coaches and had won at least 60 percent of their matches that year, and a group of 33 wrestlers from the same teams who were rated as average or below average. Discriminate function analyses failed to demonstrate differences for the criterion groups.

Also, the wrestlers differed significantly from established norms only on the tough-mindness factor. On the other hand, Morgan administered the Eysenck Personality Inventory to English-speaking wrestlers who participated in the 1966 world tournament, and he reported that performance in the tournament was significantly correlated with the extroversion dimension. Success, however, was not related to neuroticism. More recently, Morgan and Costill administered the EPI, the IPAT Anxiety Scale, and the Depression Adjective Check List to a group of experienced marathoners.

Correlations were computed between performance in the marathon, the maximal oxygen intake of these subjects, and the psychologic variables. Marathon performance was not significantly correlated with any of the variables. This is somewhat surprising since most observers of the marathon readily agree that these runners represent a "special bred". They are, of course, quite unique from an anatomic and physiologic standpoint. This uniqueness, at least in part, explains the lack of significant correlations between the variables cited; that is, they are so similar that their homogeneity suppress the likelihood of significant correlations.

8

MOTOR LEARNING AND ATTENTION

Children begin to learn so soon that it is very difficult to say what is inherited and what is acquired. When speak of children leaning, we usually think of teachers teaching. During pre-school years children learn very rapidly without any organised teaching and without any subdivision of knowledge into subjects. Children are joyfully, playfully active and as a result they develop bodily skill, they gain knowledge. They learn how to adapt their behaviour to the needs of other people. All three types of learning-off bodily skill, knowledge, and appropriate social behaviour –go on at the same time. Under the influence of the creative impulse children approach their environment as a vague unanalysed whole, and they begin to explore it in all directions. Very soon after birth, the child begins to churn its legs, fling its little arms and turn its neck this way and that. In this way he begins to control his muscles. This development of physical activity is called motor development. Crow and Draw have clarified that the tern motor activities is used for the bodily actions made possible by the working of the muscles and nerve.

Factors Affecting Motor Learning

Several factors seem to affect the role of tension or

excitement in the learning of motor skills. Most investigators have reported that simple motor tasks are learned more effectively under tension. Wechsler and Hartogs reported that high levels of anxiety hinder the learning and performance of tasks involving fine motor coordination. A third factor is the induced tension's characteristics and relation to the task. Physical tension which is nearer the performing body part, appears to be most influential.

(1) High Levels of anxiety hinder the learning: Wechsler and Hartoges reported that high levels of anxiety hinder the learning and performance of tasks involving fine motor coordination. Anxiety is a state of unrealistic fear. In anxiety there is a vague feeling or general discomfort that something unpleasant will occur. Usually anxiety is centred round such psychological points as insecurity, inadequacy, loss of prestige, or loss of material possessions, or a threat to one's life. A certain degree of anxiety is normal. But a prolonged and continuous state of anxiety is the first sign of psychotic or neurotic tendency. As a result of this, the learning or performance of a task is disturbed greatly. The sport psychologists view that the better an athlete's self-concept, the lesser the chance of being anxious.

(2) Induced tension's characteristics : A third factor is the induced tension's characteristics and relation to the task. Pressure is another source of stress. It is the expectation or demand made on an individual by parents, coaches and other members of the society. Pressure thus forces the individual to activate himself beyond his capacities. Pressure may stem from internal or external sources. Internal sources centre round one's own aspirations and ego-ideals. When the athletes have a high level of aspirations, the pressure is continuous and severe. An athlete breaks down if there is wide discrepancy between aspiration and capacity. Overambitious parents/coaches set up certain high levels of achievements in various field like games, scholastic work and other extracurricular activities. These act on the individuals and drive them almost to braking points. The existence of tension seems to have varying effects on mental tasks of different degrees of complexity. A moderate amount of tension seems to aid simple mental tasks. Bills, for instance, found that subjects who were squeezing hand dynomometers in both hands were better able to learn nonsense syllables, memorize and add columns of digits and recite scrambled letters. Zartman and Cason reported that continuous tension (a foot applied to a resisting pedal) did not aid subjects in solving a complete arithmetic problem.

(3) Simple motor tasks are learned effectively under tension : Psychological stress is essential in learning simple motor tasks. Tension is commonly found in individuals. Tension has negative as well as positive value. It is an opportunity when it offers potential gain. An example will make it clear. The superior performance that an athlete gives in "clutch" situations. Such individuals often use tension positively to rise to the occasion and perform at their maximum.

Motor Development in Various Periods of Childhood and Adolescence

(1) Childhood : As the body grows more and more,

the child has to learn a variety of new activities. At this stage the child's body grows and develops. This is a process in which the child faces many difficulties regarding balance and adjustment. Children between the ages of six and nine have a reasonably good mastery over the basic motor skills of sitting, standing, walking, running, throwing, jumping, climbing, pulling, pushing, rolling etc. In this phase more movements are acquired e.g.; catching, hopping, skipping etc. Improving muscle coordination helps the child to acquire new skills and improves the already acquired one's. Thus the six and seven year olds may be seen trying to climb a stairway by taking two steps at one time. They also try climbing and jumping down from high places. By age twelve muscle coordination almost perfect.

During the Middle childhood phase, there is fast development of endurance. The development of endurance is limited to basic endurance. The development of speed endurance takes place at a very slow pace. During the end of the late childhood phase, there is the beginning in the process of sexual maturing. There are individual and sexual differences regarding the beginning and end of this period of human life.

During the Pubescence, speed abilities can be improved. This period often results in disturbances in motor learning ability.

The parents, teachers and the coaches should tackle pubescents sympathetically when the pubescents are feeling disturbed or frustrated.

(2) Adolescence : The individual crosses the threshold into adolescence after completing his/her childhood. By this time, the child has already developed the motor skills he needs, but the activities need to be directed towards specific goals. Many studies indicate that many vigorous activities are intimately connected with physical growth. As a result of this, children spontaneously take interest in play which involves running and jumping. Consequently, they manifest a distinct liking for football, hockey, basketball, cricket, etc. In fact, new enthusiasm begins to manifest itself in the child when he/she has reached the age of eleven or twelve. This vigour, zest or excitement is called adolescence. Strength abilities continue to develop at a fast pace during adolescence. The development of endurance is training dependent.

Adolescence id a good phase of development of coordinative abilities. Motor learning ability also improves considerably during the adolescence.

LEARNING AND PERFORMANCE

From the very beginning of behaviourism, one of the main interests of behaviourists has been the study of learning. But the concept of learning comes very close to being a mental phenomenon. Consequently, behaviourists have sought to define learning precisely in terms of observable behaviour. To begin with the carefully distinguish between performance, which is what a person or animal actually does, and learning which has to do with changes in what he can do. Some behaviourists, then, attempt to describe performance only, avoiding the use of the concept of learning. But the majority try to infer the process of learning from careful analysis of changes in an individual's performance.

Inferring changes in learning from changes in

performance is not as simple a matter as it may seem. Improvements in performance are not always a result of learning. A football player may play better this week than last week for many reasons. He may have had a cold last week, or he may have had a fight with the coach, or perhaps this week's game is especially important. And when performance changes for the worse or does not change at all, one cannot assume that something learned earlier has been forgotten or that nothing has been learned.

In the laboratory, the psychologist can study animals that he is certain have had no previous experience with the task to be learned. For example, a hungry rat can be taught to run a maze for a food reward. If it runs the maze faster on Trial 20 than on Trial 1, the experimenter can tentatively conclude that learning has taken place, because he knows the past history of the rat. Once the animal has eaten its fill, however, it may be reluctant to run the maze as fast as it did before. Its performance is impaired, but what it has learned is not—when the animal is hungry again, it will demonstrate the learned response in as lively a fashion as before. Similarly, outside the laboratory, it is safe to assume that a person who successfully negotiates an advanced ski slope has learned how to ski. But the mere fact that he knows now to ski does not mean he will ski as well as he can on all slopes at all times. He may, for example, ski better when people are watching than when they are not.

Another important goal of behaviourism, in addition to precise description of observable behaviour, has been precise description of simple experimental techniques that can be used to control behaviour. If one can specify that, under certain conditions, a particular experimental technique has a consistent effect on performance, then one does not necessarily have to be concerned about whether the performance is a result of learning. By using the technique with an animal or a person, the psychologist can produce the desired performance. In this way the psychologist can control much of behaviour by simple manipulations of the environment. B.F Skinner has been the most influential proponent of this type of behaviourism. His approach has been called psychological engineering because he is more interested in determining how to control behaviour than in devising theoretical explanations of the processes that underlie behaviour. This approach is reflected in behaviour therapy.

But even behaviourists who are not sympathetic to Skinner's approach prefer explanations of behaviour that remain as close as possible to the experimental manipulations made. They usually explain learning by relating it to two well-established experimental techniques: classical conditioning and operant conditioning.

CLASSICAL CONDITIONING

Classical conditioning involves reflex behaviour. A reflex is an involuntary response that is elicited by a specific stimulus. For example, a sharp puff of air on the eye evokes a blink of the eyelid; a tap on the patellar tendon produces the knee jerk; the prick of a thorn causes reflexive withdrawal of the hand or foot. In classical conditioning, a new stimulus comes to elicit the response. For example, if several times in a row a buzzer is sounded just before a puff of air is blown at your eye, you will probably blink the next time your hear the buzzer, whether the puff of air follow of not.

The noted Russian physiologist I. P. Pavlov discovered classical conditioning in the course of a series of studies on the physiology of digestion, for which he won the Noble Prize in 1903. Pavlov was originally concerned with the preparatory reflexes of the stomach—that is, how the stomach prepares itself, by secreting digestive juices, for food that is placed in the mouth.

He found that the mouth also prepares itself, by secreting saliva, for food that is merely seen or smelled. This is a learned response; when an animal first sees or smells an unfamiliar type of food, it does not salivate. The fundamentals of conditioning that Pavlov unravelled in studying the conditions under which an animal learns to secrete saliva have had an extraordinary influence on nearly every theoretical position in the psychology of learning.

Strictly speaking, however, Pavlov was not the first to study the conditional reflex. That honour goes to a little-known American psychologist named E.B. Twitmyer. About the turn of the century, Twitmyer was a graduate student at the University of Pennsylvania. For his dissertation experiment he chose to work with the patellar reflex, or knee jerk. Twitmyer rigged up a small hammer that, when released, would strike the subject's patellar tendon. As a warning that he was about to release the hammer, he always sounded a bell beforehand. One day while he was testing a subject, he accidentally hit the bell. Although the hammer did not drop, the subject's knee jerked anyway. When Twitmyer asked him why, the subject said he didn't knee jerked anyway. Weh Twimyer asked him why, the subject said he didn't know—the response was involuntary. Twitmyer, recognizing that he had stumbled onto something very interesting, abandoned the original plan for his dissertation in order to study this fascinating new type of learned response. Although he reported his findings at the annual meeting of the American Psychological Association in 1904 none of the scientists there saw the significance of his work. In fact, it was the first published description of what we now call classical, or respondent, conditioning. Twitmyer, disappointed in the reception the scientific world gave his experiments, abandoned them entirely.

Within a year or so, Pavlov made essentially the same discovery working with dogs that Twitmyer had made with human beings. The scientific world gave Pavlov little encouragement at first either, but he was an established investigator—a Noble Prize winner with his own well-equipped laboratory—and he persevered. Finally other scientists saw the importance of his work. So Pavlov's name, not Twitmyer's is associated with the discovery. Because his work was more complete and more elegant than Twitmyer's we shall discuss it in detail.

The salivary reflex, like all other reflexes, consists of a stimulus—food in the dog's month—that elicits a response—salivation. Pavlov discovered that, if a previously natural stimulus itself gradually comes to elicit salivation. In Pavlov's terms, the food in the mouth is an unconditional stimulus. It elicits the unconditional response of salivation. The term "unconditional" is used because the response does not depend on the organism's previous experience with the stimulus. In other words, an unconditional reflex is innate, not learned. The new stimulus that comes to elicit salivation is called the conditional stimulus. The salivation that it evokes is called the conditional response, or conditional reflex—conditional upon the animal's previous experience. In other words, a conditional response is learned. Any one of a large variety of stimuli can serve as a conditional response is learned. Any one of a large variety of stimuli can serve as conditional stimulus for salivation: the sight of food, the sound of a metronome or a bell, the flash of a light.

ATTENTION

Modern psychologists consider attention as a mental process. Attention exists only when someone is attention to something. It is better to say, "I attend to my work" then "My attention is directed to my work." Attention is an activity of the person attending. An athlete may not attend to what the coach says or does in the training class. Such an athlete may be termed as inattentive. A coaching camp to be effective must receive attention of the sportspersons. Inattentiveness of the athletes is a problem to the coach. Attention is the selective activity of consciousness. Inattention is not lack of attention. Attention and inattention are not opposed mental processes. The athletes are said to be inattentive, when they are not attending to any area as desired by the coach but they are attending to something else. When the athletes are attentive, they are attending to the area/ object desired by the coach. For an example, if our tendency to consider too many thoughts at the same time does not match the attentional demands of a situation, the performance is likely to suffer.

The process of concentrating the mind upon a particular activity with a view to achieving a specific object is called attention. Psychologists describe it as the focusing of attention upon a particular object. When an athlete wants to participate in the athletic meet, he pays attention to practice sessions fully. This process of concentration of attention has been accepted as a major factor in learning. Whatever the object on which our attention is focused, our sense organs and motor organs adapt themselves to take part in the activity required. As soon as the athletes present themselves in the coaching class, their ears, eyes, necks and their bodily positions adapt themselves with reference to the coach.

Sports psychologists speak of training athlete's powers of attention and concentration. They believe that athletes would do well in any sport activity if they really paid attention. Attention makes for greater efficiency, understanding and intellectual achievement.

Attention is selective. It facilitates to some stimuli and inhibits responses to others. It is a preparatory stage for effective activity. It indicates a degree of readiness on the part of the individual. That is why coaches put much premium on attention and attach great importance to incentives. These incentives may induce athletes to give their best attention.

Selectivity relates to how a person limits attention to selected objects or ideas. These are mentioned below:

- Can you pick out the open player for a pass?
- Can you pick up on cues to your opponent's movements?

Alertness relates to the keenness of cognitive attention. These are given below :

- Are you alert in paying attention for the starting

signal?

- Can you maintain your concentration at the end of a long and-living endurance event?

Sport psychologists are beginning to examine attentional styles and the relationship of individual differences in attentional style to sport performance and behaviour.

Types of Attention : We may distinguish three kinds of attention.

1. Involuntary (Random)
2. Voluntary (Non-(Volitional)
3. Volitional (Habitual)

The first kind is known as Involuntary or random attention. This is characteristically attention with that effort and is involuntary. This is the most primitive form of attention. This kind of attention is not sustained for any length of time because as soon as any new stimulus occurs, the attention shifts. An example will make it clear. Rajesh is keenly listening to the cricket commentary broadcast by AIR. All of a sudden he hears shouting by a group of people on the streets. He looks outside for a moment and again continues listening to the commentary. Rajesh's mother reminds him of his next day's examination and asks him to study. Rajesh hesitates and tries to read unwillingly, but soon he tries to the cricket commentary. This is forced upon us hence it is involuntary or random type of attention. When Rajesh pays attention to the loud noise on the streets, it is forced on him. Such attention does not last long.

The second kind of attention is Voluntary Attention. Whenever we intentionally look or listen, attending is referred to as voluntary. It is usually imposed on the individual by some such motives as reward or punishment. This kind of attention is predominantly unpleasant, random and unnatural. An example will make it clear. Rekha does not like to read her history book, but she has to study it for the sake of the examination. Her attention is not spontaneous, but requires considerable effort or will. She has to control the distractions.

The third kind of attention is Non-volitional Attention. It is spontaneous. It develops on account of real interest in the thing itself. This kind of attention is effortless. There is no feeling of strain. It is a fruitful kind of attention. The individual is well set to follow the chosen course of activity with interest. In the case of Rajesh, we have seen above that because of his interest in cricket, he is absorbed in listening to the commentary.

Distractibility in Attention

Distraction may be defined as any stimulus whose presence interferes with the process of attention. We feel disturbed when we have an experience of an itching nose, an ill fitting collar, a broken finger nail, the hum off conversation. The lay mind classifies distraction as a condition causing bad workmanship. But experimental evidence yields conflicting results. In some cases, the distraction is deleterious. In others however, paradoxical results are obtained, in that better work is performed during distracting influences.

The effect of distraction on different individuals is a very varied one. Some people find it very difficult to overcome even sight disturbances in their environment.

The least noise is sufficient to draw their attention from the task in hand. For others, the converse appears to be true. It is also a matter of common experience that get used to distraction. An example will make it clear. A student living by the side of a railway station in Bombay gets used to the noise of the local trains; it is not a distraction for him. Surprisingly enough, he may be disturbed if the noise suddenly stops. Internal distractions such as anger, fear, boredom and worries have a more powerful effect than every day external distractions.

Distraction can be overcome in a number of ways. The most common way is to put more energy in the task itself. This can be done by creating more interest in the task. If this is not possible the next best thing is to increase motivation by artificial means, such as reward or bonus. Another method of overcoming the effect of distraction is to build up habits of inattention to the distracting stimuli. This is the case with those persons who live near a railway station or on a very busy roadside. The noise and whistle of the moving train does not affect their peace and tranquility. This process is known as the process of negative adaptation.

Strategies to Develop Attention

We attend to objects which interest us. Interest is described as the "subjective" condition of attention and leads to it. This description is misleading because it is not possible to say which comes first. The two factors seem inseparable. We attend to objects in which we are interested. We are interested in objects to which we attend. These two aspects develop together. Attention is objective in the sense that it is concerned with some object other than the process of attending. Interest seems to be subjective, interest in the person himself. Secondly, attention is more fluctuating than interest. Interest is enduring and persistent.

Interest has been described as the feeling-tone of experience and activity. This does not mean that we always attend to what is pleasant. Interest is not identical with pleasure. We are often impelled to attend to the unpleasant. It is the feeling of pleasure or displeasure which accompany the urge to attend. Avoiding danger, seeking food, looking after the young ones and the like. These are basic to life and impel us to attend to certain things, and are a source of interest. Thus, it may be argued that things are interesting because we are impelled to attend to them and not that we attend to them because they are interesting.

If we attend to things which are interesting and turn away what is unpleasant and uninteresting. How is it that we attend to things which are uninteresting and unpleasing? An individual thinks of some end or aim which he/she wishes to achieve. When he/she is able to think of ends, he/she has to think of means to that end. Thus, even if means are not interesting or pleasant he/she continues to attend to them for the sake of the end they are likely to help him/her to achieve.

Dimensions of Attention

Nideffer proposes that attainment style exists along two dimensions, width and direction. Width ranges from narrow to broad. Narrow attention is focusing on a limited range of cues whereas a broad focus on one's own thoughts and feelings to an external focus on objects or events outside the body. Let us look at Nideffer's model of attentional focus.

Nideffer felt that varying combinations of attentional width and directions are appropriate for varying sport activities and that the use of the appropriate attentional focus can enhance performance. A broad internal focus is an analytical style that is useful for planning strategy and analysing previous performances, coaches often needs a broad internal style. A broad external focus involves taking in a good deal of information. Many sport situations call for a broad external focus. An example will make it clear. A basketball player is leading a fast break and a football quarter back is trying to pick out primary and secondary receivers. A narrow external focus is useful for activities that require concentration on a ball or object, such as bowling, batting in basketball/soft ball, putting in golf and kicking a football. A narrow internal focus is appropriate for mentally rehearsing a task or performing strength and endurance activities such as distance running and weight lifting.

Nideffer developed the Test of Attentional and Interpersonal Style (TAIS). This test assesses individual's tendencies to use broad internal and broad external information and to narrow attention effectively. This test is also useful in assessing an individual's tendencies to become overloaded with internal or external information.

9

RESEARCH METHODS IN SPORTS PSYCHOLOGY

As a social and biological science, psychology is considered the new school or discipline on the block primarily because a century of scholarship is not significant when compared to the discipline of philosophy, chemistry, mathematics, and astronomy, which have existed for centuries. While noted scholars such as Socrates, Plato, Desecrates, Michelangelo, Galileo, Newton, and Copernicus can trace their affiliation and contributions to scientific disciplines that have existed for centuries, the reputations of noted scholars in psychology and sport psychology are equally deserving.

Sigmund Freud, B.F. Skinner, James Watson, Carl Rogers, and Eric Erickson are recognized names in the world of psychology; these psychology made a tremendous contribution to the application and scientific study of behaviour. While psychology as the parent discipline can identify with scholars who, at the very mention of their name, elicit admiration and respect, sport psychology has a proud and still-emerging history that is representative of outstanding scholars who have dedicated their professional lives to the area.

Landers, Bird, Harris, Martens, Gould, Kozar, Lord, Miller, Cratty, Kroll, Fisher, Singer, Griffith, Morgan,

Nideffer, and Tutko and Ogilvie in their early work are recognized for both their significant teaching and research in sport psychology, basis and applied. Where did it all begin and what has shaped the content of sport psychology? The intent of this section is to provide a brief historical overview of sport psychology in the United States.

The rationale for this limiting geographical representation is not ethnocentric or provincial in nature but rather one of experdiency and relevancy. It should be understood by the reader that within the parameters of international athletic competition between the world's superpowers, America's nemesis, the Soviet Union, has been extensively involved in the research, study, and application of psychology of sport in human performance since before the Russian Revolution of 1917.

The German Democratic Republic, Poland, and Bulgaria as well as Great Britain, Canada, Cuba, Italy, and France have demonstrated considerable research accomplishments in sport psychology. These countries receive research funding from government sources as well as generate research in their respective colleges and universities. It is important to note that within the America scientific community there is general consensus that a gap exists between the acquisition of new theoretical knowledge and the time that theory is put into practice.

Most of the leading scientific communities in other nations do not have a wide a gap as exists in the United States. However, stringent government control in the United States relative to the use of human subject research could account for the scientific gap between theory and practice. Although information from Soviet Union and other Eastern bloc countries in regard to sport psychology is very limited, the reader is encouraged to examine the following, but by no means comprehensive, sources for current basic and applied research in sport psychology: Journal of Sport Behaviour; Perceptual and Motor Skills; Journal of Sport Psychology; Yessis Review of Soviet Literature; International Journal of Sport Psychology; Sex Roles; Dissertation Abstracts International; and Completed Research in Health, Physical Education, and Recreation.

Although the first psychological research specific to sport in the United States was conducted by Triplett in 1897, the honour of being America's first sport psychologist belongs to Professor Coleman R. Griffith. As the first American to intensively investigate the formal study of the psychological aspects of sport, Professor Griffith began his investigation in 1918 at the University of Illinios.

His research efforts, initially directed to some of the psychological factors involved in basketball and football, attracted the attention of the director of athletics. George Huff. It is evident that Huff was somewhat of a visionary in regard to recognizing the enormous impact sport psychology would eventually have. Professor Griffith was a profile scholar. After receiving his Ph.D. from the University of Illinois in 1920, he stayed at the University until his retirement in 1953.

In 1938 his athletic research laboratory was closed; from 1944 to 1953 Professor Griffith was provost of the university. In one of Griffith's first experiments, he utilized a Sanborn Reaction Timer to measure the reaction time of football players at the Illinosis Athletic Field. According the Griffith:

> Some attempt was made to divide men into groups according to their "quickness", but the formal measurements were liberally interpreted and modified by what the coaches already knew of the physical and mental traits of the man under examination. The one fact that appeared independent of these "interpretations" came during the second year, when a rather significant difference in the reaction time of the men's two squads seemed to be correlated with pronounced difference in the success attained during the two seasons. That is, the squad of men with the shorter reaction times seemed to be more successful than the squad with the longer reaction times.
>
> Further studies and observations were made in vision and attention, the former with a view to helping...to make a more skilful use of the whole field of vision in sizing up a situation or play, and the latter in selecting those types of formations and shifts which offered to the opposing team the most confusing spread of apprehension or span of attention.

It seems rather ironic that Griffith's work and interest in studying and analysing the field of vision in relation to constructing shifts and formations in order to confuse and exasperate the opposition over a half century ago is remarkably similar in design and purpose to the multiple and complex shifts and formations utilized by the National Football League's Dallas Cowboys. It is of interest to note that several national football league teams have utilized, for quite some time, sport psychology and psychometrics.

Dr. Griffith taught a special section of that traditional and most often required "Introduction to Psychology" course that almost all students take; however, this special section was reserved strictly for athletes. In the fall of 1923, this course was specifically modified and changed into a new course titled "Psychology of Athletics". The approach of "Psychology of Athletics" was to examine and analyse psychological data and theories by specifically utilizing the athletic environment rather than general society.

The content of "Psychology of Athletics" included attention span, emotional, vision, reaction time, and reviewing current literature that was relative to the psychology of sport. From a pedagogical position, Dr. Griffith believed it necessary to present and analyse various problems and methods that coaches deal with on a regular basis. An example provided by Griffith relative to coaching methods of the day addressed the "effect of fear and related emotional conditions upon the rate of learning and upon the retention of habits."

The end result was a book titled Psychology of Coaching, a series of observations on what the various coaches at the University of Illinois did to prepare their teams. In 1925, George Huff approached Professor Griffith with a plan for establishing an athletic research laboratory which "would devote itself to problems in psychology and physiology of athletic activity quite independently of any attempt to create bigger and better athletic teams".

The laboratory was established in 1925 and located in the newly completed men's gymnasium. According to Griffith, "so far as we have been able to determine, the Illinois laboratory is the second psychological laboratory devoted wholly to problems in athletics. There is, at the Coaching school in Berlin, Germany, a psychological laboratory in connection with a half dozen other laboratories and clinics...." Griffith's laboratory consisted of one room specifically designed for use as a workshop, animal laboratory, and physiology lab. A rat colony from the Wistar Institute was used by Griffith for learning experiments, exercise effects, and similar research studies.

PROSPECTS FOR THE FUTURE

When analysing the historical evolution of sport psychology, one could conclude that the research and applied nature of the discipline has, and may well continue to be, multidirectional and multidimensional. Some may feel, that because of the various directions in terms of methodology and theory that have been tried and disregarded over the years, sport psychology is germinated and in a state of chaos, while others will undoubtedly view this situation as intellectually healthy and stimulating.

However, it would appear unlikely that the professionals interested in and involved with sport psychology as a field of study would ever say the historical evolution and future prospects are severely limited. An innovative and yet seemingly obvious proposal is the construction of a new and potentially dynamic paradigm in sport psychology that was presented by Bird and Ross who proposed that physical educators develop theories indigenous to their own discipline and interest and not borrow from the parent discipline. From both a pure research and applied perspective, their proposal is timely. Some psychologists engaged in research in human performance appear to have little tolerance or respect for physical education professionals regarding their knowledge base of psychology in general and human behaviour in particular and, again, there may be merit in their assessment.

On the other side of the coin, there are coaches, athletes, and other physical education professionals who seem to be uncomfortable with psychologists in terms of what psychologists claim they can do to enhance athletic performance; there is an enormous difference between "knowing about" and "knowing how". For the most part, coaches and athletes may not be too receptive to a clinical psychologist who arrives on the field and insists he or she can help both coaches and athletes especially if the sport psychologist does not have a background in competitive athletics.

THEORY DEVELOPMENT

Physical education is a unique and important academic discipline in terms of methodology and mission; as a scholarly area of research and presentation, physical education is both interdisciplinary and intradisciplinary. In the same mode that exercise physiology utilizes the research methods of the parent discipline of psychology as opposed to finding a permanent home in physical education.

Professionals in both academic areas, physical eduction and psychology, have chosen sport psychology as their primary research and teaching interest. One of the result of this multipronged approach to the investigation of sport psychology, in historical terms, is the development of appropriate methodologies with which to actually study and research the area and the construction and application of a sound theoretical basis.

Are both physical education and psychological theories equally utilized? Psychological theory has been the dominant force primarily because of theory development. Theories and methodologies have changed over the years, as reflected in the different thrusts that sport psychology has undertaken. It should be noted that upon the demise of Professor Griffith's athletic research laboratory in 1938, interest and research in sport psychology in the United States became dormant for a number of years. Bird and Ross and Landers provide an analysis of the theory building and development in sport psychology.

According to Landers, the development of a theoretical basis for sport psychology, commonly referred to as theory testing, can be divided into three periods. During the period between 1950 and 1965, the majority of research was oriented toward developing an empirical approach in order to assess personality types.

In regard to this initial thrust, Bird and Ross state that "typical comparisons were those made between groups such as athletes and non-athletes or elite and beginning participants in the same sport. The assessment instruments employed were generally standardized tests of a multivariate nature. The basis of test selection was availability rather than theoretically based".

The information deduced from this type of data

collection method was at best inconclusive, and personality testing remains an intriguing but still inconclusive area for sport psychologists. Because of the unsatisfactory results of personality testing and the recognized lack of theoretical approach, the majority of sport psychology research deviated from personality testing between the mid 1960s and mid 1970s and proceeded to move toward one of two major research thrusts—theory testing and Interactionism.

Landers as reported by Bird and Ross states, "Theory testing, as it was conceived at the time, generally involved attempts at testing existing theories from the parent discipline to determine their value in providing a theoretical basis for sport psychology and motor performance. One of the premises being that once workable theories from the parent discipline were determined appropriate for basic research in sport psychology and motor behaviour, the application of the findings would become part of the tools and techniques of applied sport psychology.

The primary examples of such theory borrowing were drive theory and allied studies which tested the social facilitation effects as they were thought to affect motor skill acquisition. The overall results generated on the basis of such endeavours were rather discouraging.... As a result, similar to the empirical atheoretical framework which preceded it, the laboratory oriented, theory based approach also fell out of scientific favour." During this same period, Landers identified Interactionism as the other, and still continuing, form of research.

After examination, analysis, and operational

research efforts in atheoretical personality testing and the subsequent but brief interest in theory borrowing, sport psychology researchers turned "their attention to the task of attempting to select and test one or two psychological factors which seemed to be systematically related to... performance".

Interactionism attempts to study and determine the extent, if any, that one or two psychological factors such as anxiety or self-concept have upon performance in a specific setting. Settings and all their attendant environmental contents are thought to have a direct influence on the performance of an individual relative to the psychological state of the individual.

For instance, a great basketball player with constant high anxiety and low self-esteem who is about to compete for the league title on a neutral court or hostile court could be a research topic of Interactionism. What is the perceived value of the results and intent of Interactionism relative to sport psychology or motor-behaviour? According to Bird and Ross, "Much worthwhile evidence is still being conducted from the intercationist point of view. However, the results of the vast majority of those studies have allowed for neither descriptions of motor performance effects or greater insights into individuals' relevant perceptions of situations. They have not allowed us to explain what it is that anxiety does within the person that cause differential performance as a function of identifiable levels of anxiety".

In concert with the recent shift from the theoretical to the applied has been "the tendency to disfavour the laboratory in favour of the field and an associated decrease in attempts at theory building specific to motor skill performance". In tune with the times, Professor John Silva at the University of North Carolina has formed the Association for the Advancement of Applied Social Psychology, which holds promise for both the researcher and the practitioner.

10

GROWTH AND DEVELOPMENT

Physical and motor qualities are very much influenced by the growth and development of the children. The book written by Espenschade and Eckert on motor performance from mental development through infancy, childhood, and adolescence to adult hood and old age. A volume on physical activities as related to growth, development and on health of children and youth, growing knowledge base in motor development by Rasick are considered to be the milestones in the field of growth and development.

GROWTH STUDIES

Review of educational research contains much references regarding growth and development. Growth studies not only concentrated on physical growth patterns but also pertains to social, psychological and mental parameters. Even the longitudinal growth studies were also conducted by the physical educationists. Some of them have been given below :—

MEDFORD OREGON BOY'S GROWTH STUDY

Medford Boy's growth study[1] was conducted by the Physical Education Research Laboratory of University of Oregon from 1950 to 1968. The purpose of the study was,

(1) To construct physical and motor mean growth and growth acceleration curves for boys of 7 to 18 years of age and to study the nature, scope and significance of individual differences of tests and representing body structure, muscular strength, muscular endurance, muscular power, ability and reaction time.

(2) To relate these physical and motor growth factors to physiological maturity, physique type, nutritional status, socio-personal adjustment, interests, scholastic aptitude and achievement.

(3) To identify those physical and motor factors that are the most significant growth indicators, particularly as related to adolescence.

(4) To contrast all traits in the study for boys who make and who do not make athletic teams competing in inter school competition, to make the same contrast for boys who score high and low on strength tests and batteries, agility and speed, muscular power and reaction time.

(5) To revise and construct strength and other tests for boys of all ages and to determine the interrelationships of the factors included in the study of various ages.

100 boys were studied in the first year at the ages 7, 9, 12 and 15 years and 40 boys at ages 10, 11, 13 and 14 years. In the subsequent years the first group (group of 100 subjects) were retested until they graduate, and 4 types of analysis were made as mentioned below —

(a) Cross-sectional with 40 boys at each age from 9 to 15 years inclusive.

(b) Longitudinal with boys followed up until high school graduation.

(c) Convergence at the end of 4 years in which boys in the longitudinal phases were joined at the over lapping years of 9, 12 and 15 years.

(d) Single year in order to hold constant, the effect of chronological age on those variables that change with age.

Except the scholastic aptitude, and achievement test the remaining all the tests were converted on the subjects within 2 months of their birthdays, in order to assume reasonable homogenity as related to chronological age. The following items were included in the test –

(1) **Maturity factors :—** Chronological age, skeletal age and pubescent development.

(2) **Physique type :—** Sheldon's Somato type components.

(3) **Body size measures :—** Body weight, standing and sitting heights, leg length, hip width, girth of upper arm, chest, abdomen, buttocks, thigh, calf and lung capacity combination of Wetzel Grid and Skinfold measures over triceps, apex of scapula, faternal abdomen, these measures were also used as indices.

(4) **Strength elements and batteries :—** Grip strength, back lift, leg lift, pull-ups, bar push-ups, Roger's strength and physical fitness indices, 11-cable-tension strength tests of muscular groups throughout the body.

(5) Motor tests :— 60 yard dash, standing broad-jump, total body reaction and movement times.

(6) Scholastic aptitude and achievement :— Grade point average and the following tests depending upon age. California mental maturity, otis quick-scoring Gamma standford achievement, Gates, reading-Iowa test of educational development.

(7) Psycho-personal adjustment :— The following tests depending upon age, sociometric questionnaire, Cowell personal distance scale, Cowell social behaviour trend, Adjective check list, California psychological inventory, Mental health analysis, level of aspiration.

(8) Interest :— The following items depending upon age, childrens interest blank, adjective check list. Dreese and Mooney Interest of inventory, what I like to do, Kudor performance Record, Garretson and Symonds interest form, Strong vocational interest blank.

MATURITY

Maturity plays an important role on the performances in physical activities. The more matured individual in the same age group will have more advantage. Maturity in the growth studies can be assessed by skeletal age and pubescent assessment rather than the chronological age.

(1) SKELETAL AGE

Skeletal age is a common indicator of maturity and this can be determined through X-rays. Each and every bone of an individual begins as a primary centre of ossification, and passes through various stages of calcification enlargement and shaping till the individual reaches adult status. These changes can be seen in an X-ray, which distinguishes ossified areas from the areas of cartilage, where ossification has not yet occurred. The wrist area is very common in practice.

Greulich and Pyle[3] have presented an atlas containing hand-wrist X-ray reproductions separately for boys and girls. In assessing skeltal age some 30 bones and epiphyses of an obtained X-ray were compared with those in the atlas. The average of all assessments is taken as the individuals skeletal age which is usually given in months.

In the Medford study conducted by Clarke, boys were tested within 2 months of their birthdays at each age from 7 through 17 years. The standard deviations fluctuated between 11.9 and 15.1 months. The ranges varied from a low of 58 months at 17 years and a high of 85 months at age 13 years. The median range was 71 months i.e., nearly 6 years. The magnitude of these differences at 14 years of age was that, approximately $\frac{2}{3}$ of 168 boys tested had skeletal ages between 12 to 16 years, but these boys did not very more than 4 months in the chronological age. The skeletal age range for these 14 year old boys was much greater i.e., between 10.3 and 16.9 years. The results were, more mature boys were taller, heavier, and stronger and had a greater potential for success in interschoolastic athletics, they were prone to higher levels of aspiration and better psychological adjustment.

PUBESCENT ASSESSMENT

Pubic assessment is used to evaluate the maturity

of adodescent boys and girls. Normally, the growth of pubic hair is considered for this, and also in boys, the characteristics like deeping of voice and increase in the development of genetic organ will appear. In case of girls the age of manacher (first monthly course) will be considered for the maturity.

Clarke and Deguits[4] compared the skeletal ages and various physical and motor factors with the pubescent development of 10, 13 and 16 year old boys.

Physical maturity was differentiated by pubescent assessment most effectively at 13 years of age, although it was not so sensitive to maturational changes as was skeletal age. Even though, the distribution by pubescent assessment of 13 year old boys disclosed some boys at this age in each of the 5 pubescent catagories, most of them were equally distributed pubescent groups 2 and 3 of the classification categories suggested by Greulich[5] as mentioned below —

1. Prepubescence
2. Appearance of fine pubic hair
3. Pigmentation
4. Appearance of kinky or curly hair
5. Post pubescence

At 10 years of age, nearly all boys were classified in group 1, at age 16 they appeared mostly in group 5 with some in group 4. Over lapping were appeared in other groups. 13 and 16 year old boys who were advanced in pubescent development, for their respective ages had higher mean scores on body size, strength and motor tests. At 16 years of age greater percentage of ectomorphs were formed in group 4 farther than in group 5.

PHYSIQUE TYPE

The type of physique or the body type is having much importance in the field of growth and development as it is having influence on the performance and behaviour.

Kretschmer has suggested the below mentioned terms—

(1) Pyknic — To represent the fat body type.

(2) Athletic — To designate the muscular individual

(3) Aesthenic — To represent the lean body type.

Sheldon and others has classified the body types into 3 main categories as mentioned below :—

(1) Endomorphy — Fat body, predominance of soft roundness throughout the various regions of the body.

(2) Mesomorphy — Heavy, hard and rectangular physique with rugged massive muscles, and large prominent bones - Athletic type.

(3) Ectomorphy — Frail, delicate body structure with thin segments. Lean personality.

Further, they have stated that all the individuals will be having some qualities of all the three components. Hence he has suggested as follows :—

(1) Endomorphy - (7-1-1).

(2) Mesomorphic endomorph - (621, 631, 642, 721, 532).

(3) Endomorphic mesomorph - (261, 361, 462, 362, 352).

(4) Mesomorph - _1-7-1).

(5) Ectomorphic mesomorph - (162, 163, 264, 263, 153, 253).

(6) Mesomorphic ectomorph - (136, 236, 245, 126, 135).

(7) Ectomorph - (1-1-7).

The numerals in the brackets represents the amount of components the first numeral represents endomorphic quality, the second one mesomorphic and the third ectomorphic. This classification is also called as Somato Typing.

The following generalisations can be made from the analysis of Medford study :

(1) The endomorphic boys are handicapped by excessive body bulk, by low strength relative to size, by inability to perform arm-shoulder muscular endurance tests and inability to make and be successful in the games and sports competitions. They have negative self-images, but seen psychologically well adjusted.

(2) Mesomorphic boys have advantages related to greater gross and relative muscular strength, muscular endurance, successful participation in games and sports, peer status and psychological adjustment.

(3) Ectomorphic boys have disadvantages related to less body bulk and gross strength, but they have advantages related to relative strength and arm-shoulder muscular endurance. They show some lack of peer status and psychological adjustment and some favourable self-image characteristics.

Skeletal age has significant but low correlations with Somato type correlations.

ANTHROPOMETRY

Anthropometry is the oldest type of body measurements, it is an early type of testing in physical education. Hitchcock and Sarget produced profile charts to study how the individuals can be compared with their standards. Sargents chart consists of 44 anthropometric measurements as well as a number of strength tests.

The great use of anthropometry is to determine relationships between body structure and motor performance.

The another use of these measure is to show the succession of annual means during the growth period. It is very difficult in deciding the site (area) and obtain a precise measures.

The instruments that are used in the authropometric studies are very few and quite simple. Normally stadiometer is used to measure the standing as well as sitting height, the breadth measures are obtained by use of hiding wooden or metal calipers, circumference of limbs are taken by a metallic or steel tape and the skin folds are determined by fat calipers.

SKINFOLDS

Measurement of skin fold thickness is commonly used technique for estimating total body fat. The sites generally accepted for the determination of body fat are - (i) at the back of the arm, (ii) at the subscapular position on the back and, (iii) on the mid-axillary line at the level of the umbilicus.

From the studies conducted on skinfolds, the total amount of fat may be determined as the appropriate percentage of body weight.

MUSCULAR STRENGTH

The muscular strength is defined as the maximum contractive power of muscles. It is normally measured with a dynamometer or a tension meter. The tests using dynamometer includes right and left grips, back and leg lifts. The tensiometer tests are used in testing the strength of 38 muscle groups throughout the body.

Muscular strength is very much related to the entire physical motor complex of individuals, including their maturity, physique, body size and motor and athletic abilities. Gross strength shows high performance, stronger boys will have better psychological development, greater peer status and more self-image.

MUSCULAR ENDURANCE

Muscular endurance is defined as the ability of muscles to perform work, or lifting a constant load for longer time. Muscular endurance to exist must possess muscular strength also to some content. For e.g., to perform the pull ups, an individual must have sufficient strength also. Some times, muscle groups of the same strength may also differ in endurance.

Hence, the seperation of effect of growth and development is very difficult, because increase in the muscular strength affects the muscular endurance also.

MOTOR PERFORMANCE

Motor performance includes a wide variety of physical tasks such as running, jumping, throwing, swimming, diving, gymnastics etc. Hence study of such capacities is of great importance in the studies pertaining to growth and development of children.

Many tests are being conducted to determine the motor performance. They are Listed below :—

(i) Standing broad-jump

(ii) 50 yard/30 yard sprints

(iii) 60 yard shuttle run

(iv) Vertical jump

(v) Target throw

(vii) Soft ball distance throw

(vii) Velocity over hand base ball throw etc.

Only two of them have been explained below :—

STANDING—BROAD—JUMP

It is nothing but the jumping horizontally from a standing position. It is the indication of explosive power of the leg muscles.

Clarke and Deguits[7] studied the relationship between standing broad jump distance and various maturity, structural and strength measures of 12 year old boys. The correlations obtained were as follows :—

(a) Anthropometry - .41 with body weight, leg length and lung capacity.

(b) Cable-tension strength - .52 with elbow flexion and hip extension.

(c) Combined variables - .69 with elbow flexion strength, body weight, hip extension-strength, ankle plantar flexion strength and leg length.

(2) 60-yard Shuttle run

The 60 yard shuttle run is used as a measure of speed and agility. Gates and sheffield[8] experimented agility run with 18 tests, out of which 15 were involved change of direction while running and 3 were motor ability elements. In this study, 60 yard shuttle run was considered a good indicator of speed and agility.

To overcome the problems in administering the shuttle run, a common running surface should be provided.

11

PSYCHOLOGY OF COUNSELLING AND COACHING

COUNSELLING

There is found contradictory views about use and importance of counselling in area of sports activities. The concept of counselling may create anxiety in many sport coaches who contend that they know very little about counselling others, yet successful coaches counsel their players all the time, some more effectively than the others. Although counselling is a learned skill, practised by licensed psychologists, coaches are still able to use fundamental counselling skills to enhance communication with players.

Although it is known that coaches are not expert in performing function of counselling, but role of counselor should be integrated to the coach's responsibility. It is important to remember that coaches and anyone else without the proper credentials, are not registered psychologists, because of which cannot be referred to themselves as psychologists, and must not engage in counselling process in areas related to the athlete's psychological well-being. There are many issues that players bring to the sport venue that require the aid of a trained counseling psychologist, but it is not possible to make arrangement of such professional counsellors all the time, because of which it becomes necessary for the coach to have knowledge of when to refer the players to a person trained in clinical psychology.

To understand that how a coach performs function of counsellor and what is the need to provide counselling to players, it is first necessary to discuss the meaning of concept of counselling, which is as follows :—

Meaning of Counselling

The tool or means by which normal people are provided help to cope with normal problems and opportunities is known as counselling. Counselling is a full fledged process, which can be long-term sometimes in which responding to attitudes, feelings and behaviours of others can be included. It is also very necessary to accept the perception of human being and feelings without personal prejudice and outside evaluation standards. It implies that if the coach and players engaged in a honest and meaningful dialogue, the coach must accept player's feelings and perceptions as real rather than denying that the player's feelings are legitimate.

Even without professional training, coaches can become better counsellors. Process of counselling is quite similar to process of interviewing, in which coaches keep on engage all the time. Generally, following tasks are included in functions performed by counsellor :—

- gathering information so that he can become a good listener,

- helping the players to solve problems by looking at alternative approaches and solutions to issues, and

- giving information or advice.

Whether it is necessary to seek advice or help of professional, it depends to a lot of extent on severity of problem. Issues in which it becomes necessary for the coach usually relate to player's participation on the team, however, issues that are not directly related to players performance can be resolved only with discussion.

While performing the function of a counsellor, it is very necessary for the coach to keep in mind that counselling can be risky, especially for the players, if not done properly. This function requires one to expose deep-rooted and sometimes negative attitudes toward other persons. Another risk lies in being candid in comparing the player's perceptions of a situation with the coach's views. The main function of the coach is to help players to feel secure about disclosing personal feelings and perceptions without risking their self-esteem or participation on the team. Such disclosures may benefit the players on emotional grounds and can also help them in improving their performance level.

Need to Counsel Individual Players and Teams

A lot of stress exist in all competitive activities. Thus, although sports activities are called only a game, but participation in an organized competitive sport is stress-inducing. Often players need assistance in the areas of time management and study skills. It is found that various players do not obtain desirable grades because they have not been trained to manage the time constraints placed on them. Such kind of problems may inhibit concentration, sport performance etc.

Coaches should counsel players about and provide resources for achieving good performance. The coach should be empathetic about daily pressures and time constraints with which players are confronted. Sometimes players set their own lofty and unrealistic goals. This is unfortunate, because fulfilling the dream of competing as a professional player in any sport is quite rare. It is very necessary to prepare the players psychologically as well as physically for a life that is independent of participation in competitive sport is obvious. This is an important function performed by coach.

As known that on basis of number of players participating in competition, sports activities are divided into individual and team activities. Winning is a team effort and so is losing. If team members consistently need to resolve issues among themselves, they will have less energy and concentration available for competition. Observable coaches ask various kinds of questions from players.

It is very important for the coaches to know or analyse the level of satisfaction of all the players involved in team. An unhappy player is an unmotivated and ineffective one. Team counselling is necessary for responding to personal problems or problems with interpersonal relationships. A team's performance depends on some specific characteristics, level of morale is one among them. For players to feel high group satisfaction and to support one another mutually is desirable. When morale is not at preferred levels, then counselling with all or part of the team to gain insight into the issues that causes team dissensions is in order. Coaches cannot dictate heightened cohesion, morale and team member satisfaction. Then, the main objective of team counselling becomes to facilitate player discussion on an individual or group basis to solve problems or to confront and deal with various kinds of problematic issues.

Important Guidelines for Effective Counselling

It is not necessary that only licensed or professional persons can help other persons in solving out their problems. It has been proved by studies that people who are often in a position to hear the concerns and feelings of others can be quite effective in the role of confidant. It is necessary for a coach to know the art of counselling, for which he should have some specific qualities in him, which are as follows :—

a. Listening : Habit of listening to other's views is known as attending behaviours in professional terms. In this behaviour, various techniques are included, which facilitate the client's need to talk. This is considered to be the most critical aspect of counselling process. Non-professionals should sit back and listen attentively and actively to the views of other person.

It has been proved that effective coaches are active listeners. They maintain constant consistent eye contact, face the speaker squarely with a slight forward trunk lean, and use encouraging gestures. Not only this, they avoid distracting movements and avoid cutting off the speaker in mid-sentence or mid-thought, which allow the person to continue expressing his feelings.

b. Be Supportive : Giving support is a response phase, while showing concern is a part of listening phase of communication. It does not take a degree in counselling to show empathy and understanding toward the player's feelings. It is fine to disagree with the player's point of view on a matter, but do not make him feel inadequate as a person for having a different perception or opinion. Support can also mean helping the players to face facts.

c. Showing Concern : Regardless of any discrepancy between coach and player's perceptions of situation, it is necessary that coach take player's feelings seriously. If the player feels that the issue is serious, then it is serious, no matter what anyone else think. It is the player's perception that are important. Coaches who ignore or do not show sincere concern about helping the player cannot expect to receive the player's loyalty. Disloyal players do not tend to contribute to the team effort.

d. Respecting Privacy : It is only in privacy that conversations between coaches and players should be conducted. This applies intensely if competition is contentious and the tone is argumentative. Privacy also means confidentiality. The player's feelings are no one else's business. Thus, it is very necessary for the coach to maintain a kind of privacy in dealing with players.

Concept of Short-Term Team Counselling

It is necessary that term related unresolved issues

must be surfaced and be confronted with the proper attention of all concerned parties. Otherwise, the player's full and proper attention and concentration will be diminished and re-directed away from game preparation and skill development. Team counselling consist of meeting in relatively small groups to share concerns, to explore common problems and to gain insight knowledge. Attempts are made to meet the objectives of the sessions in relatively short order so that issues can be quickly resolved and team members can give their proper concentration to preparing for the competition.

In team counselling, individual or personal issues are not given much importance, however, one exception of this rule is the situation when a player is unable to establish a positive relationship with any other team member, an issue that affects the attitudes or performances of others.

Different Approaches to Counselling

For identifying and using specific counselling skills, an important approach which is generally used or taken into account is microskills approach. This technique of counselling is based on the assumption that effective counselling requires certain techniques that lie within broader types of skills. These broader skills are attending behaviour and influencing behaviours. Influencing behaviours involve the coach in causing growth or change in the players, while attending behaviours have three dimensions, namely eye contact, appropriate body language and verbal following. It is very necessary to discuss both of these strategies properly, which is as follows :—

Strategies of Influencing Behaviour

In this technique, it is important to listen and to react verbally. Influencing behaviour is a function of the coach's making the right verbal responses. The counselling skill used for this strategy is called responding, which are of two kinds, namely, responding to content and responding to feelings.

Responding to Content : It implies that coach communicates his understanding of player's experience. It involves imagining another person's world as though the coach is the player. Coaches can best facilitate this empathetic understanding by encouraging the players to be specific about experiences or feelings, and by not indicating personal attitudes the player. Players should feel that they can explore their feelings without fear of retaliation. Responding to content also involves the coach's ability to show the player that he understands the player's situation or feelings. This is done by carefully listening to the details presented by the player and rephrasing the player's expression in a fresh and new manner.

Responding to Feelings : There are different ways in which players can express their feelings. They may do this directly through their choice of words or indirectly by tone of voice, by describing a situation without stating their feelings etc. Players who walk in a dejected manner with head down or who isolate themselves from other after being taken out of the contest might also be making a strong non-verbal statement about their feelings of disappointment. Whatever the form of expression, the coach's role is to respond to the speaker's underlying feelings.

Strategies of Attending Behaviour

Not every coach is comfortable with the task of discussing another person's feelings or with discussing his own. Coaches who want to have nothing to do with counselling athletes in any form, those who resist listening and verbally responding to player's feelings, should avoid this role completely. Do not fake it. Players can clearly tell when a person is insincere, uncaring or uncomfortable, however, it is very important that another responsible adult maintain this role. Players need someone with whom they can talk, and if possible, who is in an influential position with the team.

From various studies, it has been found that counselling behaviours can be classified into three categories, namely, under-participation, over-participation and distracting participation. Under-participators are considered as incompetent as well as insecure. They convey to other persons the unwillingness or an inability to offer assistance. In a sport setting, this lessens the player's faith in the coach. The typical underlying cause of under-participation may be the coach's fear of involvement either with the player or a particular issue. The counselor withdraws in both verbal and non-verbal directions.

In case of over-participators, they may try to cover up their anxiety, as evident in the person who always has to be the centre of attention and who is constantly talking. This is description of an insecure person who is afraid to lose control of the environment or who needs attention to reinforce his self-worth and to overcome self-doubts. In reality, over-participating a way of exerting control, which reduces level of anxiety. Usually it is seen that over-participatory coaches confront the players and jump to conclusions with little awareness of the player's feelings. This kind of situation also occurs when the counsellor is doing most the talking or does not let the person finish his statements.

Third kind of participator, namely, distracting participator over-responds to irrelevant issues, sometimes with loud laughter. The issues may never be addressed because the conversation becomes bogged down in trivia. Coach should make pleasant comments, as this tends to reduce nervousness and defensiveness. Only then, he should address the gut or complex issues. An important tendency found in distracting participator is to shift issues hastily before the primary issue has been resolved or at least dealt with at length. A frequent change in issues tell the player that the main issue has not been heard or is not being taken seriously by the coach. The counselling coach should try to get minute details relating to issue and should try to stay with it until a sense of closure is felt by player.

To foster openness and honesty, it is necessary for the coach to make use of appropriate counselling technique. Some of the strategies which are generally used by coaches for this purpose are following :—

- Facial Animation : In this strategy, coach show interest and active mental involvement in conversation through various facial expressions.

- Occasional head nodding : By this strategy, coach can reinforce the player's commitment to talk effectively. It also represents that the coach is interested in the issue.

- Good eye contact : If coach does not keep direct eye to eye contact with the player, then a feeling of distrust can be aroused in him. Actually, avoiding eye to eye contact and glancing everywhere, but not at the player show disinterest.

- Occasional Smiling : While listening player's view, it is necessary that coach keep on smiling from time to time as this relaxes him and makes the coach appear compassionate, understanding and human.

- Soft tone of Voice : Effective communicators avoid appearing defensive and, instead, remain in control and relaxed, which fosters honesty and continued communication.

- Gestures of Hand : With this, speaker can keep the listener's attention intact. Use of hand gestures should be done properly.

COACHING

There are two major areas in which psychology can make a significant contribution to coaching and sport:

1. maximizing the athlete's performance

2. maximizing the athlete's personal growth and development

It is, however, becoming increasingly apparent that there is an extremely important relationship between the two areas. In order to make significant and consistent progress in maximizing performance plenty of attention must be paid to the athlete's personal growth and development. The psychology stresses that athletes have to contend with today are often tremendous and the vast majority of athletes who consistently perform well are those that "have their head together". They are the ones who are psychologically mature, have stable personalities, and whose personal growth and development have not been neglected.

They are the ones who keep things in perspective and cope well with difficult situations because they are "cool", adaptable, independent, self-confident, and self-disciplined as well as committed. It is important to think about how kids get that way and whether there are things you can do to help develop these kind of characteristics. They certainly don't happen automatically and the practical suggestions presented later in this paper can be a big help. In the end it is the athletes who have to be able to cope, react, and perform on their own.

In order for them to effectively do so it is critical that the coach pay particular attention to the personal growth and development of his/her athletes.

Planning with the Athletes

Prior to conducing the planning sessions with the athletes the coach should have done considerable preparation and homework. A fairly comprehensive list of areas of concern should have been developed along with examples of specific goals and possible strategies. The coach also should have listed any limitations within which the group has to operate that season. The group should be made aware of limited practice time, required dress code or any other organizational, institutional, or economic limitations, prior to any goal setting or detailed planning.

Clarifying these issues early results in more realistic goal setting and avoids the frustration of these things coming up later and disrupting the group process. The group planning sessions should revolve around the same questions the coach has been considering:

1. What are some "Specific goals" to be strieved for in each area?

2. What is the season or activity really all about?

3. What the some plans or "Strategies" to achieve those goals?

4. What are "areas or concern" for the team?

Many athletes have never been asked to contribute to a process like this before and the homework the coach has done is a big help in stimulating discussion and getting contributing. It may be necessary to ask questions of different individuals and suggest possible areas of concern. It may be necessary to give examples of specific goals and possible strategies.

It may be necessary to give goal setting and planning "homework" exercises to designated group members or simply allow time between sessions for thought about the topics. It is however important to set a co-operative motivational climate at this stage and have as many ideas as possible originate from the group. The more the group contributes to the plan the more likely they are to feel committed to it. The coach is being democratic at this stage but not "laissez-faire" and it is important to be organized and appear confident, goal-oriented, and committed.

The athletes should feel free and functional in contribution to the discussion but recognize the rights of majority or consensus and be aware of the commitment necessary to achieve goals. Although some coaches seem to feat the consequences of letting athletes contribute to goal setting and planning, this author found in a variety of settings that group consistently set tougher more desirable goals than one might have thought.

The coach's biggest task is usually keeping the goals and plans realistic, specific and achievable. It is important whenever possible to state goals or objective in specific, measurable, behavioural terms. Push for specificity in goal setting sessions-it results in more meaningful, realistic, achievable goals. When statements of intent are specific, and explicit, the correlation with eventual behaviour is much higher because people have a better understanding, feel more responsible, and can be held more accountable.

It is important especially with team sports to push for a consensus on goals or strategies. Discussion usually results in an acceptable compromise for all. People tend to be idealistic optimistic and co-operative at the start of the year and quickly recognize the utility of coming to an agreement or consensus.

If however, someone's goals are completely incompatible with the group goals and strategies, the beginning of the year is the time to find out. This gives these individuals an opportunity to adapt or find a group whose objectives are more compatible with their own. For example, a team may agree to approximately equal playing time during playoffs. A player who has difficulty accepting this goal should find a solution before problems develop.

During all these group planning the coach or someone he designates should jot down or record any specific goals or strategies about which there is a consensus. A list can be compiled and developed into a comprehensive seasonal plan which can be viewed and distributed for final approval so that everyone is clear about group expectations and the commitment required.

Planning by the Coach

The coach should start by simply thinking about what the season or activity is all about and what his/her goals, priorities and expectations are for the group. It then might be beneficial to start thinking about the upcoming season in three ways.

First, what are the possible areas of concern for the group this year. Secondly, what are some of the specific goals that can be set in each of the areas. And thirdly, what might the strategies be in pursuing these goals. The coach should also start to think about how the athletes can be involved in this goal setting and planning activity.

It is important to find out that why the athletes came out and what they are expecting as soon as possible. If the coach's goals, priorities and expectations are incompatible with those of the athletes the preseason is the time to find out so that both sides can find an acceptable solution while options still exists. There are also other reasons for involving the athletes in preseason goal setting and planning.

Psychologists have gathered evidence to suggest that

involving the group in discussion and decisions can increase the awareness of and the commitment to group goals. Involving the group in goal setting also results in a from of "psychological contracting" which can be an extremely effective motivational and leadership technique. At any rate it is advisable to plan and conduct a number of group planning sessions at the beginning of the year for the purpose of involving the athletes in creating and compiling a comprehensive list of goals and intentions.

Making the Plans Work

The coach's role obviously is to encourage behaviour which is consistent with group gaols and discourage that which isn't. However, as mentioned earlier, if goals and strategies are clear individuals soon start to monitor their own behaviour and that of their team mates in a very consistent way. In the early stages if leader behaviour is not completely self-evident the coach should explain his/her actions and rationally tie them in with stated goals and strategies.

Whenever possible the coach should concentrate on recognizing behaviour which is desirable and consistent with group plans and praise it openly while explaining why. Also do not restrict praise to certain individuals- find something positive in everyone's behaviour and let them know about it! It is extremely critical that a coach's behaviour is consistent with the stated goals and strategies.

Youngsters particularly will usually model what adults do rather than what they say. Therefore, it is ridiculous to demand sportsmanship, poise and discipline, etc., from one's players yet regularly get overexcited, abuse officials, etc. It is important to be aware of one's own behaviour and its effect on others.

If coaches are to expect commitment from their players they must display that commitment and dedication themselves! One of the best ways to reinforce commitment is to involve all group members in forms of leadership. Contributing to leadership produces a "vested interest" or increased identity with the groups as well as an increased appreciation of the leadership role and group plans. With all little guidance and ingenuity group members can become a terrific source of helpful ideas and leadership.

When goals and strategies are clear athletes can easily be asked to create and conduct drills, "fun" games, and activities. This provides the coach with the opportunity to give special attention to individuals or subgroups. Different members can be asked to act as subgroup leaders or captains for each game, practice, or team activity. Participant Leadership can be a great source of enthusiasm and group morale! A periodical review of stated goals and strategies serves as a reminder or restatement of intentions and commitment.

In additional it provides measurable and observable feedback to the group. Areas of success become source of pride, reinforcement, and inspiration. Areas of failure become the focal point for effort and concentration. A review also provides an opportunity to reassess goals and intention and modify or adjust them if deemed necessary.

Inrtermediate goals can be developed in areas of difficulty and extended goals in areas of success. It might be pointed out that when detailed group planning has been done the coach usually makes fairly regular reference to stated goals and strategies in the everyday justifying, reinforcing, or criticizing of behaviours in the group. The coach can also do a number of things to try and insure that other factors in the athlete's environment encourage commitment to the group plan. For example copies of stated goals, intentions, and priorities can be made available to "significant others" in the social world of the group.

This will often increase the commitment of group members as a result of the sincere public display of intentions. It also usually produces considerable moral support and helps educate parents and supporters. The coach can also try to insure that the total competition or participation structure encourages the kinds of behaviour being strived for. This may be difficult and frustrating if the existing league or competitive structure as a whole contains behavioural problems!

These planning and leadership techniques usually result in tremendous personal growth and development in young athletes. In addition, the much more complete preparation inevitably produces dramatic improvement in physical performances. Specifically, the following things tend to happen:

Clarified goals and priorities—everyone has a better idea as to what the season or activity is all about and why it is important.

Increased commitment and motivation towards group goals.

Improvement in the athlete's self-confidence and in group morale.

Fairly immediate success because many specific achievable goals have been identified.

The development of psychological maturity in athletes which is reflected in self-discipline and skills in self-control and self-management.

Increased appreactional of the important of planning and goal setting in any activity.

The elimination and prevention of problem behaviours.

Increased poise and improved abilities to adapt, cope, "keep things in perspective," and concentrate on the right things.

Increased empathy and respect for the rights and feelings of others.

Improved communication and understanding among all parties concerned and control over "unintentional" problem behaviour.

Happier athletes, better performance, and much more fun.

The Gymnast's Perception of the Coach

One of the various techniques utilized in the data collection phase of current work was that of a questionnaire specifically developed for competitive gymnast. The questionnaire has been slightly modified to accommodate the particular experience of European gymnasts and to facilitate translation in the future. Basically, however, the questionnaire seeks the same material wherever it would be administered. Participants in filling it out was voluntary, as were all aspects of the work, and the gymnasts were told either in a covering letter or in person that they could ignore items they did not wish to answer.

As one can imagine, there is an enormous wealth of material that has resulted to date from this one approach, but in this article we will deal with one particular question and an analysis of the answers received.

First of all, the qualities specified by responding gymnasts were given in two ways. Some athletes answered in a prose form, and it was necessary to translate this into a list. This was accomplished by simply taking the characteristics as they were given and assigning a number to them, e.g., 1, 2, 3, etc. as they appeared in the answer.

The majority of responses gave the qualities in a numbered order, and it as concerned. We have also made the assumption that the coaching characteristics given reflected a deeply felt need in the gymnast and indicated his or her feelings about the important of certain attributes in the coach. The overall pattern that emerged in response to this question clearly showed a trend suggesting that the gymnast's view of what was important in a coach bore a close relationship to their own level of performance competence.

It was evident that the more competent gymnasts were naming a somewhat different set of qualities in their trainer than were their less successful colleagues. The further down the success ladder, the more scattered the expressed attributes became, except that in the less proficient athletes technical knowledge was almost always identified as very critical. Technical knowledge was also given a relatively high priority by top performance of a technical nature. It was rarely the most important characteristic identified by them, in contrast to the less competent performer, for whom it was ingredient sought.

There appeared to be an inverse relationship in this regard, that is, the less successful the performer, the higher significance the coach's pure technical ability became, and the more successful the competitor, the less crucial that quality was based on its placement and description in the list of trainer characteristics. For the top performers the major characteristics listed were the general area of psychological motivation and support.

The higher skilled the individual gymnast, the more important were these kinds of elusive characteristics in the coach. The more skilled and successful gymnasts appear to psychologically need and seek greater personal interactive closeness and motivational incentive based on an emotionally supportive relationship. Although the less competent by gymnasts also need "motivation", it was not specified in incorporated terms but in very task-oriented concrete language, such as "a coach who wants me to do well", or "who wants to win" versus "a coach who will help me realize my potential". Variation was present in the lists complied for both highly successful and less successful gymnasts.

However, the degree of variation was considerably higher in the athletes whose actual performance record was not indicative of current competence. Further data analysis revealed the fact that for the most part the better athletes always listed the same specific qualities among the top characteristics named. Although they varied in placement, they were always present near the head of the lists.

As previously indicated, in the vast majority of cases the number one characteristic sought by top performers was the coach's willingness and desire to get to know the gymnast as an individual and in a way that suggested care and emotional support. The verbal expression of this quality varied in construction, for example, "a coach interested in me as a person", "a trainer who would respect and know me as an individual," "a person I could trust and who cared about me," but the characteristics all suggested the same general theme. This kind of coach-quality was rarely indicated as of highest significance in the lists of less successful gymnasts.

For that group, as discussed earlier, the major quality sought and identified was technical knowledge. This was indicated in such phrases as "a coach who knows how to teach difficult skills," "a person with knowledge of the mechanics of tricks," "an individual who is able to show me how to do various moves". Other high-regard characteristics named by the more competent gymnasts were the ability to motivate and maintain motivation, the ability to manage the sociology of a team, the ability to develop meaningful interpersonal relationships on a one-to-one basis as well as a group basis.

In the less successful gymnasts more concert attributes such as physical strength, skilled spotting, and the provision of discipline were seen as more critical and more often mentioned. It is interesting to note that two characteristics having high valence for all gymnasts were the ability to organize training efforts and a clearly shown love for the sport. It may be of interest to our readers to know what some of the other characteristics were that were identified by gymnasts.

Although, some of these qualities might appear as behaviour objectives many thoughtful coaches strive to develop, it is encouraging to find them confirmed by the athletes themselves. These are given with no attention to order of importance as they appeared on the gymnasts lists, but they do represent the most frequently mentioned attributes identified by the must successful youngsters.

In addition to the emotional-social factors, the capacity to communicate technical knowledge on an individual basis, a genuine dedication to gymnastics, and organizational ability, the primary qualities discussed were: the ability to get a concept across with an appreciation for the individual gymnast's learning style; a minimum of excess verbiage consistent limited setting and the firm establishment of authority; an applied sense of humour while still taking gymnastics seriously; the absence of sarcasm and negativism; as a general mode of response; hardwork and enthusiasm on the part of the trainer; personal predictability, fairness in dealings with gymnasts; mature to relating good control of individuals and team at competitions; the ability to translate theory into action; honest critique of work being done and not being afraid to say the truth about one's performance; the ability to make a firm decision without dwelling on it; exemplary behaviour the ability to say calm under stress; capacity to allow others to take responsibility; the ability to direct the spirit, not break it; openness to suggestions and willingness to admit error; willingness to provide the opportunity for gymnasts to socialize with one another and to listen to the gymnast's concerns and problems; reliable attention to safety factors; and on overall positive attitude about training and competition.

The above listing is not exhaustive, and there were others, including some totally unpredictable and even bizarre notions, but the characteristics given represent a reasonable summation of the majority of responses. There are several rather interesting issues raised by the material reported herein. It would be most enlightening to see if the task-oriented characteristics identified by less successful gymnasts at a given point in their careers shift to the more emotional-caring ones as their own level of ability improves.

Another related intriguing inquiry would be to follow the less successful gymnasts who, at the time the questionnaires was completed, did not fit the pattern associated with their level but in fact named characteristics more in harmony with those expressed by better gymnasts, to see whether they would become more successful themselves over time. In essence, which comes first, the chicken or the egg, the success or the particular coaching approach associated with it? What is really going on here? Is there some specific psychological cluster reflected in a deeply held conviction and statement concerning the nature of the teaching-learning relationship which is predetermined in the personalities of successful athletes? If this is so, and if it isn't present in a particular gymnast, can it be acquired or developed? This is not just a matter of intellectual curiosity, for it has implications concerning the training of gymnasts. The material reported in this article strongly suggests that homogeneous ability grouping of gymnasts is a wise way to construct programs.

Although it is true that some positive benefits from the performance of more skilled gymnasts might "rub off" on less competent members of a group, our results suggest that the more highly functioning youngsters need a different type of coaching style and approach which warrants segregation. Unless the coach is Renaissance man or woman who can easily shift technique moment-to-moment depending on whom he or she is working with, it is far more efficient to form groups whose members are very close to one another in ability. This may appear like an obvious fact which anyone could figure out without much fanfare.

Experience suggests, however, that when such grouping does occur it is haphazard or done for overtly valid but not thoroughly understand reasons, such as "They are easier to work with" or "They require less spotting." The essential question is, why is this the case and what does it say about the kind of training interaction needed? Our gymnastics history replete with examples of how we do the correct thing but for a less than informed reason. It may seem to some that it is a subtle difference indeed between a coaching approach to a group or individual that says, "All right, I want you to do 20 compulsories each and watch execution on the cartwheel" and the coach who says, "Let's see where we need to look at this exercise for perfection; what would you say we have to focus on at this point?" But, as a matter of fact, these represent one hell of a difference in both mentally and approach. It is important to note that many of the better performers come from well-established program where several coaches and combinations of coaches are employed.

Very often the coaching responsibilities are divided along performance level lines, one coach responsible for beginners, another for intermediates, etc. Obviously, from material reported here and elsewhere, such training organization makes eminent sense. Coach selection, however, should not be random or simply based on seniority but should be made on the basis of who can provide the kind and quality of input and leadership needed at a specific point in the gymnast's development.

In most cases, the more experienced coach ends up working with the top youngsters. As has ben pointed out, it is not only experience that is important, although it often accompanies the ability to motivate, but what is needed is a special personality and instruction style as well. The relationship sought by top performers is sometimes already present, particularly if the coach and athlete have been together a number of development years during which time this interaction has emerged quite naturally.

Many top youngsters were reporting on the kind of coaching they were receiving already, but this was not the case in all situations, and from comments made on questionnaires it was apparent that sometimes the characteristics mentioned were those desired but not necessarily present in their current training experience. It is at these junctions that progress may come to a standstill. This is the time when "driving" is still needed, but to a substantially lesser degree than an emphatic, personalized system of motivation and communication. Some coaches simply don't have the skill. They continue to rely on the old, faithful, standby techniques, tried and true, that worked at a prior time, failing to recognize that their gymnasts have growth and that the older methodology is no longer appropriate or effective.

At that level of competence much support is gained from comparable peers, and the depth of talent and self-determination is such that it wins out over a less than ideal coaching situation. Loyal gymnasts "hang in," but their gymnastics becomes a maintenance operation at best, rather than a creative, forward surge at a time which often coincides with the peak learning and performance years.

How much better the athlete could do should the coaching conditions be more favourable to the development of excellence remains an unknown. Perhaps this is one reason we see a considerable amount of team changing at the higher levels in this country. If we accept the hypothesis that the more skilled gymnasts need a different applied coaching style, one which is more highly personalized and motivational in nature, as suggested by this paper, it follows that we need to pay careful attention to the background, level, and kind of expertise brought to our training camps at the national levels.

Coaching staffs that work at this level should be well balanced. Not only is it important to have most technically competent coaches present, but equally, if not more critically, it is important to have coaches with a demonstrated capacity to relate to and motivate gymnasts in an emotionally supportive and psychologically caring manner. The coaches must be able to do this with gymnasts other than their own. It would seem appropriate to consider the development of a master blueprint for the training and evaluation of coaches on a national level along the lines suggested by our observations.

Currently, time in the sport is often a major criterion for selecting training camp personnel. Although this factor may be correlated with coaching ability, longevity does not necessarily ensure the present of the specific characteristics named in this research. Another common, and perhaps more reliable criterion, is to nominate the coaching staff from among those coaches who have the top gymnasts, the assumption being that if they have the best youngsters they must be the best coaches.

Again, this may be a very erroneous assumption. There may well be some coaches working in our programs today who have no top competitor in the current gymnastic ranks but who nevertheless are superb trainers with excellent style-shifting and relationship-building ability. Conversely, it is quite possible that some coaches with current top gymnasts may not be as sensitive and appropriate at a more finely tuned level as it may appear on the surface. Giftedness along the dimensions suggested in this report is not as mysterious as it may seem; the desired characteristics can be identified and should be prerequisites for selection.

Some individuals will resist this notion, perhaps for selfish reasons, but somewhere along the line soon priority will have to be given to the overall welfare of the gymnasts above any other factors or adult motivations.

In this same regard it is interesting to contemplate what would happen if we asked our top performers, all of whom know most of the coaches working on the national scene, whom they would like to see at the training camp. If this were done anonymously there is considerable evidence that some gymnasts would not endorse their own current coach. Without recourse to a secret ballot this could be accomplished by asking each gymnast to nominate two coaches, one from their own gym and one from another.

Use of Groups to Improve Athletic Performance

As part of the Athlete Development Workshop program for AAU synchronized swimming teams sponsored by the Sports Committee, groups were introduced to discuss coach-athlete and athlete-athlete relationships. It was one component of a psychologically-based regimen designed to assist athletes in coping with the results of stress in high level competition.

Other modules include application of biofeedback, use of relaxation techniques and autogenic procedures to improve performance levels....Conducted in 1977 and 1978, these groups have been utilized with the top 25 United States AAU synchronized swimming teams during the week-long workshops conducted at the Olympic Training Centre, Squaw Valley, California, as well as with a speed swimming team in a large metropolitan area. Each group consisted of the coach and, in some cases, an assistant and his or her swimmers meeting with the facilitator to discuss concerns about competition practice, performance, team goals, and mutual expectations. Usually there were 8 to 10 swimmers in each group plus the coach.

GROUP STRUCTURE AND PROCESS

During the late 60s and early 70s when the encounter group movement flourished, the typical social entree was not "What astrological sign are you?" but more relevantly then "What group are you in now?" Psychologically and growth oriented groups were given a great deal of bad press, much of it deserved. Anyone who lived through that ear is likely to have a wide range of reactions and images when presented with the genetic term group.

Unfortunately many of the images may be negative, may trigger off memories of less than positive experiences with a group led by an unprofessional, poorly trained, confrontation group facilitator. The term may elicit associations like: therapy group, encounter group, confrontation group, rap group, support group, experimental group, wide-open group, counselling group, "touchy-feelie,' or sensitivity training. On the other hand, the word may have a positive valence, connoting pleasant memories of a fruitful group experience resulting in growth or change.

Such reactions, whether negative or pleasant, can result in a perception about groups which as the potential to significantly alter any future group experience. Our expectations are powerful determinants in influencing subsequent exposure to a perceived similar experience. Because of this,...I made a deliberate attempt to demystify these perceptions about groups in advance at a general meeting of coaches. A brief introduction was given at the initial orientation meeting for athletes as well.

Heavy emphasis was given to the central purpose of the group an opportunity to discuss common concerns, goals, worries, and experiences related to high-level competition. Here was a chance also for the team to meet under more relaxed conditions without the usual time pressure of a tight weekly schedule. Following this description, they were told what the groups are not.

They are not counselling, therapy, or encounter groups. All participants were reassured that embarrassing personal questions would not be asked, their family history would not be probed, and that all questions would be answered as directly and honestly as possible by me. Informal follow-up at meals and during breaks gave me the opportunity to answer further questions of the coaches and athletes.

These contacts resulted in stronger rapport and several to lower any further resistance to group participation. At the beginning of each group discussion, the purpose and goals of the group were once again repeated, the structure explained, and the demystification process gone through.

Feedback from coaches and swimmers indicated that this was not redundant or overemphasized, rather it was seen as valuable and, in some cases, essential to the group's direction and success. The next step in beginning the group was for me to make a personal statement which led into the first phase of discussion. The responses were informative and enlightening to me and often revealing to the swimmers and coaches.

Many themes developed which were common to every group, while others were idiosyncratic. In analysing the answers given to this global question, 27 themes developed. Due to the overlap it might be more accurate to say there were consistently a dozen themes which seemed to arise in each group.

If the discussion on level of commitment was slow in beginning, related questions were introduced, such as "What are the rewards and payoffs for you in synchrony swimming?" and "Why are you willing to sacrifice precious time you could spend studying or on other pleasures?"

DEVELOPING GROUND RULES FOR ASSISTING ONE ANOTHER

Depending on the issues raised spontaneously in discussion, this topic would be approached next. What is universally generated in discussion of this topic is an awareness of strong individual preferences for giving and receiving feedback on performance. It was my observation that this was concrete and specific enough, as well as universal to all participants' experienced, that many areas were tapped. In fact, if there were no time constraints, this subject could be explored for the full two hours. Specific examples of ground rules will be given later. More attention will also be given to what swimmers way coaches do or say that is helpful to them.

This can be potentially the most explosive subject of all those mentioned. Likewise it can result in healing wounds which have severely obstructed communication for an entire team. Many times current unresolved conflicts are alluded to indirectly or are raised directly by coach or athlete in another context, such as decision-making.

Other times past misunderstandings which still are simmering are brought into the open. Again, if the subject is not brought up, the leader will refer to some other reference with a statement like "When you referred to your difference of opinion with I wonderful how you handle disagreements and conflicts on the team". As the facilitator response indicates, the attempt is to depersonalize the conflict, that is, treat it as a normal, inevitable occurrence on any team.

The message is that it is a predictable event and there are effective ways to handle disagreements. If it is clear that an unresolved conflict exists that is having a present negative effect on the team or individual, it may be briefly dealt with as an example of a conflict and how it might be dealt with.

In at least three instances, it was my judgement that more time was necessary than was available to sufficiently explore the issue. In those cases a second session was arranged for another time during the workshop. Each of the preceding sections which have been introduced will be given more detailed attention at another point in this article along with examples and anecdotes.

CONTRACTING

The summary also led naturally into an exploration of contracting. Although seldom referred to as contracting, the notion was frequently raised in the context of decision-making. The principle of giving and receiving feedback were then broadened to encompass team issues such as practice times, planning routines, determining A and B team members, solo and duet partners, and the selecting of travel squads. Principles of contracting are an extension of those laid down on giving assistance and feedback.

An interesting array of issues was typically raised for potential contracts, from such seemingly picky items as how soon before a competitive meet costumes should be worn, to more fundamental matters of how music for a routine will be chosen.

One of the first contracts worked out in the group dealt with what to a non-swimmer might seem an inconsequential problem; yet, to the swimmers competing on a highly ranked team, was cause for consistent pre-competition tension on the team.

The tension often resulted in the flaring of tempers and distraction which, for some, led to a lower level of performance than the individual might attain under calmer conditions. The source of irritation was a habit one team member had of not getting her hat and combs in place until a few moments before the team formally began competition.

For the rest of the team the norm was to have this taken care of 15 or 30 minutes beforehand. Nagging by other team mates or coaches resulted only in more attention being paid to this annoying habit and less to essential last-minute preparations. A standoff had resulted which was the source of much unproductive gossip, name-calling, and accusations.

After both the identified norm-breaker and the team discussed the reasons for their positions, a contract was decided upon, basically a compromise which satisfied both sides in the issue. The contract specified:

1. Team members agreed in turn to make no mention of the costuming until the 10-minute deadline.

2. If the deadline was met, team members would

thank her for her compliance.

3. The recalcitrant swimmer agreed to have her hat and combs in place no later than 10 minutes before a meet.

4. If the deadline was not observed, other team members would be allowed to point out her unreadiness until she complied.

This example illustrate the principle. It deals with concrete, observable behaviour, specifics the behaviour to be changed, indicates what must be done in a positive statement rather than what must not be done, and, finally, makes the time element clear. A more complex analysis of contracting, including other principles, conditions and follow-up, did not seem appropriate for the group.

Given the shortness of time and other priorities for attention it was kept at this level. Due to the uniformly positive evaluation of the groups, almost every team agreed to continue with follow-up meetings after leaving camp. The only exceptions were those few teams of the 25 that already had regularly scheduled team meetings. Final negotiations took the form of a contract which included time, location, who shall attend, frequency and length of meetings, agenda, and designated leader. This settlement was similarly used as an illustration of contracting.

SWIMMER'S AND COACHES' MOTIVATION

Previously, an introduction was made to this topic detailing swimmer's responses. Not included in that section were coaches' answers to the question of their willingness to make such a significant contribution of their energy time and resources to the sport. On the one hand, their reasons were frequently similar to the swimmers, like the opportunity for travel, making friends, and the sense of excelling in an activity.

On the other hand, their role as coach clearly dictates other sources of payoff and rewards for them. Coaches in particular seemed to welcome the permission given by the group structure to be more personal in revealing the complex motivators which keep them involved despite very low pay and demanding schedules which complicate their lives. Almost without exception, the team members were hearing some of these reasons declared for the first time by the coach herself or himself. The impact was sometimes startling, not only on the athletes, but on the coaches, too. For the coach, answering the question led sometimes to clarification of priorities, as in, "You know I realize now that I probably would have quit before now if it weren't for my compelling desire to see syncho swimming become an Olympic sport."